The Minerals, Metals & Materials Series

Gisele Azimi · Hojong Kim
Shafiq Alam · Takanari Ouchi
Neale R. Neelameggham
Alafara Abdullahi Baba
Editors

Rare Metal Technology 2019

TMS

Springer

Editors
Gisele Azimi
University of Toronto
Toronto, ON, Canada

Hojong Kim
The Pennsylvania State University
University Park, PA, USA

Shafiq Alam
University of Saskatchewan
Saskatoon, SK, Canada

Takanari Ouchi
University of Tokyo
Tokyo, Japan

Neale R. Neelameggham
IND LLC
South Jordan, UT, USA

Alafara Abdullahi Baba
University of Ilorin
Ilorin, Nigeria

ISSN 2367-1181 ISSN 2367-1696 (electronic)
The Minerals, Metals & Materials Series
ISBN 978-3-030-05739-8 ISBN 978-3-030-05740-4 (eBook)
https://doi.org/10.1007/978-3-030-05740-4

Library of Congress Control Number: 2018964007

This Springer imprint is published by the registered company Springer Nature Switzerland AG
The registered company address is: Gewerbestrasse 11, 6330 Cham, Switzerland

Preface

Rare Metal Technology 2019 is the proceedings of the symposium on Rare Metal Extraction & Processing sponsored by the Hydrometallurgy and Electrometallurgy Committee of the TMS Extraction and Processing Division. The symposium has been organized to encompass the extraction of rare metals as well as rare extraction processing techniques used in metal production and mineral processing. This is the sixth symposium since 2014.

This symposium intends to cover research and developments in the extraction and processing of less common, rare metals that are not covered by other TMS symposia. These elements include antimony, bismuth, barium, beryllium, boron, calcium, chromium, gallium, germanium, hafnium, indium, manganese, molybdenum, platinum group metals, rare earth metals, rhenium, scandium, selenium, sodium, strontium, tantalum, tellurium, gold, and tungsten. These are rare metals of low tonnage sales compared with high tonnage metals, such as iron, copper, nickel, lead, tin, zinc, or light metals, such as aluminium, magnesium, or titanium and electronic metalloid silicon. Rare processing includes biometallurgy, hydrometallurgy, electrometallurgy, supercritical fluid extraction, as well as extraction of values from electric arc furnace (EAF) dusts, waste electrical and electronic equipment (WEEE) and less common waste streams not discussed in recycling symposia. Rare high-temperature processes include microwave heating, solar–thermal reaction synthesis, molten salt electrochemical processes, cold crucible synthesis of the rare metals, and the design of extraction equipment used in these processes as well as laboratory and pilot plant studies.

This volume covers extraction and processing techniques of platinum group metals, rare earth elements as well as other less common metals, such as lead, antimony, molybdenum, tungsten, chromium, titanium, and vanadium, including electrochemical processing, aqueous processing, biological separation, microwave heating, and supercritical extraction.

We acknowledge the efforts of the symposium organizers and proceedings editors: Gisele Azimi, Hojong Kim, Shafiq Alam, Takanari Ouchi, Neale R. Neelameggham, and Alafara Abdullahi Baba. The support from TMS staff members Matt Baker and Patricia Warren is greatly appreciated in assembling and publishing the proceedings. We sincerely thank all the authors, speakers, and participants and look forward to continued collaboration in the advancement of science and technology in the area of rare metal extraction and processing.

Gisele Azimi
Lead Organizer

Contents

About the Editors

Gisele Azimi is an Assistant Professor cross-appointed between the Departments of Chemical Engineering and Applied Chemistry and Materials Science and Engineering at the University of Toronto (UofT). She is also a registered professional engineer in Ontario. Her research program is aligned well with the "Sustainability" and "Advanced Materials and Manufacturing" research themes. In her research program, she strives for achieving a sustainable future and mitigating the adverse effects of climate change through: (1) advanced recycling and urban mining of waste electrical and electronic equipment (WEEE) utilizing innovative recycling processes based on supercritical fluids; (2) industrial solid waste reduction through waste valorization to produce strategic materials like rare earth elements; (3) development of innovative materials with unique properties with far-reaching applications in structural and energy material sectors; and (4) energy storage focusing on the development of a new generation of rechargeable batteries made of aluminium. She received her Ph.D. in the Department of Chemical Engineering and Applied Chemistry from UofT in 2010. Before returning to UofT as faculty in 2014, she completed two postdoctoral appointments at MIT in the Departments of Materials Science and Engineering and Mechanical Engineering. She has received a number of awards including Dean's Spark Professorship, Early Researcher Award, TMS Light Metals/Extraction and Processing Subject Award—Recycling, and Connaught New Researcher Award.

Hojong Kim is an Assistant Professor of Materials Science and Engineering, and Norris B. McFarlane Faculty Professor at the Pennsylvania State University. He received his B.S. from Seoul National University in South Korea in 2000 and Ph.D. at Massachusetts Institute of Technology (MIT) in 2004 both in Materials Science and Engineering. His doctoral research sought to identify the corrosion mechanisms of constructional alloys in high-temperature and high-pressure steam environments under Prof. Ronald Latanision in the Uhlig Corrosion Laboratory at MIT. After graduate research, he worked as a Senior Research Scientist at Samsung Corning Precision Glass Co. Ltd. and as a project lead to improve the process yield for TFT-LCD glass melting processes. After five years of industrial experience, he returned to MIT as a postdoctoral associate to contribute to the growing need for sustainable technology. He conducted research on high-temperature electrochemical processes, including molten oxide electrolysis for carbon-free steel production with a focus on developing inert anode materials in molten slags as well as molten salt–liquid metal batteries for large-scale energy storage.

His current research interests embrace the development of environment-friendly electrochemical processes for resource extraction/recycling, corrosion-resistant materials, and energy storage systems. He was awarded The Minerals, Metals & Materials Society (TMS) Young Leaders Professional Development Award in 2013 from the Extraction and Processing Division. In 2015, he was chosen to receive a Doctoral New Investigator Award from the American Chemical Society Petroleum Research Fund. He is currently leading efforts to separate fission products (rare earth and alkali/alkaline earth elements) from molten salt electrolytes used for recycling used nuclear fuel. He was the lead organizer of past Rare Metal Extraction & Processing Symposiums at the TMS Annual Meeting (2017–2018) and is the Chair of the Hydrometallurgy and Electrometallurgy Committee of the Extraction and Processing Division.

Shafiq Alam is an Associate Professor at the University of Saskatchewan, Canada. He is an expert in the areas of mining and mineral processing with extensive experience in industrial operations, management, engineering, design, consulting, teaching, research, and professional services. As a productive researcher, he has secured two patents and has produced over 160 peer-reviewed research publications. He is the co-editor of seven books and is an associate editor of the *International Journal of Mining, Materials and Metallurgical Engineering* (IJMMME). He is the winner of the 2014 TMS Extraction and Processing Division's Technology Award.

With extensive relevant industry experience as a registered professional engineer, he has worked on projects with many different mining industries. He is an Executive Committee Member of the Hydrometallurgy Section of the Canadian Institute of Mining, Metallurgy and Petroleum (CIM). During 2015–2017, he served as the Chair of the Hydrometallurgy and Electrometallurgy Committee of the Extraction and Processing Division (EPD) of The Minerals, Metals & Materials Society (TMS). He is a co-organizer of many symposia at the international conferences through CIM and TMS. He is one of the founding organizers of the Rare Metal Extraction & Processing Symposium at TMS.

Takanari Ouchi is a Research Associate in the Institute of Industrial Science at the University of Tokyo. He received his Ph.D. in Nano-Science and Nano-Engineering from Waseda University in 2011. In this tenure, he developed electrochemical deposition processes to fabricate metal nano-structures with both well-controlled crystallinity and uniformity at the single nano-meter scale, and demonstrated the applicability of these processes for the fabrication of bit-patterned magnetic recording media for future hard disk drives. After completing his doctoral degree, he joined MIT, where he developed liquid metal batteries, which are, in principle, bidirectional electrolysis (electro-refining) cells, for application in grid-scale energy storage. As a research scientist, he led the systematic investigation of the electrochemical properties of liquid metal electrodes in molten salt electrolytes and developed novel lithium, calcium, and sodium liquid metal batteries. Since he began work as a Research Associate at the University of Tokyo in 2017, he has developed new recycling processes for rare metals and precious metals using pyrometallurgical and electrometallurgical methods. As a member of the Hydrometallurgy and Electrometallurgy Committee of TMS, he has constantly contributed to the development of the vibrant field of metal extraction, organized technical symposia at TMS meetings, and solicited papers as a guest editor of *JOM*. He has earned several awards and honors, such as the TMS EPD Young Leaders Professional Development Award in 2015, based on his contributions to electrometallurgical processing.

Neale R. Neelameggham is "The Guru" at IND LLC, involved in international technology and management consulting in the field of critical metals and associated chemicals, thiometallurgy, energy technologies, soil biochemical reactor design, lithium-ion battery design, and agricultural uses of coal. He was a Visiting Expert at Beihang University of Aeronautics and Astronautics, Beijing, China, and a plenary speaker at the Light Metal Symposium in South Africa on the topic of low carbon dioxide emission processes for magnesium.

He has more than 38 years of expertise in magnesium production and was involved in process development of the startup company NL Magnesium through the present US Magnesium LLC, UT, until 2011. He and Brian Davis authored the ICE-JNME award-winning (2016) article "21st Century Global Anthropogenic Warming Convective Model." He is presently developing "stored renewable energy in coal" Agricoal™ for greening arid soils and has authored an e-book *Eco-stoichiometry of Anthropogenic CO2 That Returns to Earth* on a new discovery of quantification of increasing CO2 returns to Earth.

He holds 16 patents and patent applications, and has published several technical papers. He has served in the Magnesium Committee of the TMS Light Metals Division (LMD) since its inception in 2000 and chaired it in 2005, and in 2007 he was made a permanent co-organizer for the Magnesium Technology Symposium. He has been a member of the Reactive Metals Committee, Recycling Committee, and Titanium Committee, and was a Program Committee Representative for LMD.

He was the inaugural chair, when in 2008, LMD and the Extraction and Processing Division created the Energy Committee, and he has been a coeditor of the Energy Technology Symposium through the present. He received the LMD Distinguished Service Award in 2010. While he was the Chair of Hydrometallurgy and Electrometallurgy Committee, he initiated the Rare Metal Technology Symposium in 2014. He is the coeditor for the 2019 symposia on Magnesium Technology, Energy Technology, Rare Metal Technology, REWAS 2019, and Solar Cell Silicon.

Alafara Abdullahi Baba is a Professor of Analytical/ Industrial and Materials Chemistry in the Faculty of Physical Sciences, University of Ilorin, Nigeria. He holds a Ph.D. in Chemistry from the University of Ilorin in 2008. His dissertation titled "Recovery of Zinc and Lead from Sphalerite, Galena and Waste Materials by Hydrometallurgical Treatments" was judged the best in the area of Physical Sciences, University of Ilorin, Nigeria, in 2010. Until his current appointment as Head, Department of Industrial Chemistry in 2017, he was the Deputy Director of Central Research Laboratories, University of Ilorin (2014–2017). He is a Fellow of the Chemical Society of Nigeria (FCSN), a Fellow of Materials Science & Technology Society of Nigeria (FMSN), and currently a member of the Hydrometallurgy and Electrometallurgy Committee of The Minerals, Metals & Materials Society (TMS). He has a keen interest in teaching, community service, and research covering solid minerals and material processing through hydrometallurgical routes, preparation of phyllosilicates, porous and bio-ceramic materials for industrial value additions. He has more than hundred (100) publications spreading through national and international acclaimed journals of high impact and has attended many national and international workshops, conferences, and research exhibitions to launch campaigns of his research. He is the recipient of several awards and honors including 2015 MISRA AWARD of the Indian Institute of Mineral Engineers (IIME) for the best paper on Electro-/Hydro-Bio-Processing at the

IIME International Seminar on Mineral Processing Technology—2014 held at Andhra University, Visakhapatnam, India; 2015 MTN Season of Surprise Prize as Best Lecturer in the University of Ilorin, Nigeria, category; Award of Meritorious Service in recognition of immense contributions to the development of the Central Research Laboratories, University of Ilorin, Nigeria (2014–2017); and Exceptional Award in Recognition of Outstanding, Selfless Service, Contribution and Performance toward Humanitarian Services by Nigeria Association of Physical Science Students among others. He has supervised and is still supervising undergraduate and postgraduate research work and currently is serving as an external examiner to postgraduate researchers in universities within and outside Nigeria.

Part I
Rare Metals I

Feasibility of Copper Recovery from Spent Deposited Sludge of Transformer Oil (DSTO) for Industrial Applications

Alafara Abdullahi Baba, Joshua S. Ayodele, Oloduowo M. Ameen, Abdulrasaq Jimoh, Folahan A. Adekola, Abdul G. F. Alabi, Marili F. Zubair, Kuranga I. Ayinla, Abdullah S. Ibrahim, Mustapha A. Raji, Daud T. Olaoluwa, Aishat Y. Abdulkareem and Fausat T. Olasinde

Abstract Due to fluctuating crude oil prices and a widespread desire to diversify Nigeria's economy, recycling initiatives are preferred when addressing economic downfall. With increasing demands for metals and gradual depletion of non-renewable resources, there is a growing need to recover metals from secondary sources. The recycling from metal wastes is necessary as the cost of safe disposal of hazardous components is quite high compared to the amount of waste produced and limited storage capacity. Depending on the type and capacity, typical transformer oil contains 90–1485 kg of oil with a varying degree of recoverable sludge containing precious metals after degradation. In this work, leaching and solvent extraction techniques were adopted to treat spent DSTO obtained from Ayetoro III 300kVA Dis-

A. A. Baba (✉) · J. S. Ayodele · F. A. Adekola · M. F. Zubair · K. I. Ayinla · A. S. Ibrahim
M. A. Raji · D. T. Olaoluwa · A. Y. Abdulkareem · F. T. Olasinde
Department of Industrial Chemistry, University of Ilorin, P.M.B 1515, Ilorin 240003, Nigeria
e-mail: baalafara@yahoo.com

O. M. Ameen
Department of Chemistry, University of Ilorin, P.M.B 1515, Ilorin 240003, Nigeria

A. Jimoh
Department of Technical Operations, Molete Business Hub, Oluyole Industrial Estate, Ibadan, Nigeria

A. G. F. Alabi
Department of Materials and Metallurgical Engineering, University of Ilorin, P.M.B 1515, Ilorin 240003, Nigeria

A. G. F. Alabi
Department of Material Science and Engineering, Kwara State University, P.M.B 1530, Malete, Nigeria

A. Y. Abdulkareem
National Mathematical Centre, Sheda-Kwali, P.M.B 118, Abuja, Nigeria

F. T. Olasinde
Chemistry Advance Research Centre, Sheda Science and Technology Complex, FCT, P.M.B 186, Garki, Abuja, Nigeria

G. Azimi et al. (eds.), *Rare Metal Technology 2019*, The Minerals, Metals & Materials Series, https://doi.org/10.1007/978-3-030-05740-4_1

tribution Transformer having 245-kg oil weight capacity with 25 kg of recoverable combined sludge oil mixture. At optimal conditions, the leach liquor was accordingly treated and beneficiated to achieve 96.8% copper recovery efficiency.

Keywords Transformer oil · Deposited sludge · Solvent extraction · Leaching Regeneration · Copper · Cyanex®272

Introduction

For many decades and specifically since 1926, transformer oil has been a major threat to the environment due to its high content in polychlorinated biphenyls (PCBs) [1]. Globally, 30–40 billion litres of mineral oil is presently utilized in transformers and the demand will tremendously increase as time goes on [2]. The transformer oil serves as an insulator and coolant within the transformer. Consequently, both government and environmental regulatory bodies currently enforce strict environmental laws to drastically reduce both accidental and non-accidental hazards associated with mineral transformer oil utilization [3]. However, the space between the fibrous insulation coils filled up with insulating oil enhances the dielectric capacities of the insulation depending on the dielectric strength, viscosity, specific gravity and pour point indices.

The degradation or deterioration of transformer oil with oxygen takes place as soon as the equipment is energized which causes a chemical reaction to occur when the oil is exposed to mixtures of heat, oxygen, core and coil components. As the oxidation reaction occurs, mineral acids and polar compounds such as ketones, alcohols and peroxides are liberated and subsequently result in sludge formation [4]. The damage to the oil insulation in the transformer is intensified by the transformer age, in which the insulating paper applied will deteriorate continuously while the transformer oil oxidizes progressively due to the presence of oxygen in the transformer oil. In order to minimize the volume of high-voltage power within the transformer as it compromises operational reliability, the goal is to develop a new technique with higher dielectric and thermal conductivity capacities as a replacement for conventional methods being used [5]. It is therefore pertinent that oil tests should be carried out at regular intervals to ascertain its breakdown voltage as the sludge contents are made up of precious heavy metals, such as Ni, Zn, Au, Cu and Co which may be toxic, but could serve as a valuable industrial product if recovered and purified [4]. Hence, for excellent environmental, societal and economic benefits, oxidized oils can be re-applied after a thorough regeneration where all unwanted pollutants including water, acidic components and other associated gangues will be significantly reduced and potentially eliminated [6].

The enrichment of transformer oil is achieved and categorized through: filtering, purification and regeneration methods. The *filtering* process takes place at 40 °C and involves oil filtering; *purification* with a drying process commenced by heating the oil to a maximum of 90 °C; and *regeneration* by degassing achieved combining the second method with additional oxidant additives [6]. Currently, the transformer oil regeneration is a green technology and adsorption technique seems to be a key step in regeneration [7], but due to the large volume of wastes generated via adsorption [8], considerable attention is given to the development of a simple, low-waste, low-cost and eco-friendly route in the purification and regeneration. The metal extraction and purification processes of some secondary wastes including spent transformer oil can be employed by either pyro or hydrometallurgical techniques or combination of the two. However, the hydrometallurgical route has been established to be of a low-cost, less waste generation and eco-friendly technique for enriching materials with low metal contents [9–12]. Considering the aforementioned reasons, the characterization, treatment and extraction of copper from spent deposited sludge of transformer oil was treated in chloride media by the CYANEX®272 extractant in kerosene diluent. The purified copper solution obtained was further beneficiated as copper oxide, a valuable industrial copper compound with wide arrays of industrial applications such as *p*-type semiconductor indicators [13].

Materials and Method

The spent sludge of transformer oil used for this study was sourced from Ayetoro III 300 kVA, 11/0.415 kV Distribution Transformer under Baboko Business Hub, Ibadan Electricity Distribution Company, Ilorin Kwara State, Nigeria. This transformer has 245-kg oil weight capacity with 25 kg of recoverable combined sludge oil mixture [4]. All reagents used were of analytical (BDH) grade. Deionized water and doubly distilled kerosene were used in the preparation of all aqueous and organic solutions.

Physicochemical and chemical characterizations of the DSTO, including selected products at optimized leaching condition, were carried out using Buck Scientific ACCUSYS 211 Atomic absorption spectrometry (AAS), Agilent 19091S – 4HP – 5MS Gas chromatography/mass spectrometry (GC-MS), EMPY 6000 X-ray diffraction (XRD), X-ray fluorescence (XRF), and Leo 1450 with Lab6 Filament scanning electron microscopy (SEM) techniques [4, 13].

Leaching Tests

Leaching experiments were carried out in 250 ml glass reactor with a mechanical stirrer and temperature control unit. 10 g/L of the dried/processed deposited sludge was reacted independently with a hydrochloric acid solution of varying concentrations (0.1, 0.2, 0.5, 1.0 and 2.0 mol/L). The reaction time was kept constant at 120 min for each experiment. The solution in the reactor after each leaching was filtered; both residue and leachates were collected for further characterization [12]. The extent of dissolution showed 28.29, 45.22, 74.56, 92.83 and 90.21% reaction efficiencies.

However, the pregnant leach solution (PLS) collected at optimal leaching was used for the solvent extraction studies [4].

Solvent Extraction/Beneficiation Tests

Solvent extraction experiment was performed using the PLS obtained at the optimized condition and then equilibrated with an equal volume of organic solution containing a known concentration of the CYANEX®272 extractant in kerosene diluent at room temperature (27 ± 2 °C) and shaken for 25 min. Equilibrium pH was adjusted by the direct addition of concentrated HCl (0.1 mol/L) and NaOH (0.1 mol/L) solutions with HANNAH pH metre. The aqueous solutions were analyzed for metal concentrations by Buck Scientific ACCUSYS 211 atomic absorption spectrophotometer. The free acid content was determined by the titration method with pH control. The concentration of copper in the organic phase was calculated from a difference between the initial concentration in PLS and the concentration in raffinate at fixed organic/aqueous (O/A) phase ratio [14]. The extraction mechanism by Cyanex®272 was examined at optimal conditions, prior to beneficiation activities.

Results and Discussion

Characterization Investigations

The results of the characterization of the spent deposited sludge transformer oil before acid leaching are summarized in Table 1. This defines the current state of deposited spent transformer oil (DSTO) compared to ASTM standard values (Table 1).

The elemental compositions of the DSTO by EDXRF gave SiO_2 (17.07%), TiO_2 (0.37%), P_2O_5 (4.1%), Fe_2O_3 (54.53%), CaO (6.75%), MgO (0.04%), Na_2O (0.3%), K_2O (0.23), V_2O_5 (0.03%), Cr_2O_3 (0.055%), CuO (13.14%), ZnO (0.77%) and Eu_2O_3 (0.6%), respectively. However, the compositions of the PLS used for solvent extraction studies contain: 1243.4 mg/L Cu^{2+}, 789.2 mg/L Fe^{2+}/Fe^{3+}, 110.7 mg/L Ca^{2+} and 23.5 mg/L Al^{3+}. Prior solvent extraction investigations showed the iron forming major gangues in the leach liquor was masked by adjusting the pH of the solution from 1.2 to 3.6 with ammoniacal solution [15] to obtain the following results: 1239.6 mg/L Cu^{2+}, 0.6 mg/L Fe^{2+}/Fe^{3+}, 2.1 mg/L Ca^{2+}, and 1.2 mg/L Al^{3+} [4].

The GC-MS results of DSTO in Table 2 indicate the presence of compounds such as Tridecane, Heptadecane, Nonadecane, 2-Bromodecane, 2-Buten-1-one, Octadecane, 4H-1,3-Benzochoxin-4-one, Pentadecane and Butylated hydroxytoluene.

Table 1 Physicochemical properties of DSTO as compared with ASTM D1816 [4]

Property	Experimental value	ASTM D1816 STANDARD
Density	0.82 g cm^{-3}	0.55–0.89 g cm^{-3}
Viscosityl	9.7 cSt	3.0–12 cSt (max)
Flash point	140 °C	145 °C (max)
pH	4.27	5.5–8.2 (min–max)
Specific gravity	0.86	0.91 (max)
Cloud point	9 °C	7–15 °C
Water content	1.32×10^5 mg L^{-1}	30 mg L^{-1}
Free fatty acid	0.04 mg KOH q^{-1} oil	0.01–0.08 mg KOH q^{-1} oil (min–max)
Acid number	0.8 mg KOH g^{-1} oil	0.01–0.03 KOH g^{-1} oil (min–max)
Pour point	−7 °C	−8 to −6 °C (min–max)
Boiling point	220 °C	120–230 °C (min–max)

Table 2 Compositions of organic compounds present in the processed spent transformer oil (%) [4]

Compound	Yield
Tridecane ($C_{13}H_{28}$)	6.30
Heptadecane ($C_{17}H_{36}$)	10.24
Nonadecane ($C_{19}H_{40}$)	3.17
2-Bromododecane ($C_{12}H_{25}Br$)	9.14
2-Buten-1-one ($C_{13}H_{20}O$)	4.78
Tritetracontane ($C_{43}H_{88}$)	11.90
Octadecane ($C_{18}H_{38}$)	4.48
Tetradecane ($C_{14}H_{30}$)	13.38
4H-1,3-Benzochoxin-4-one ($C_{10}H_{10}O_5$)	3.66
Pentadecane ($C_{15}H_{32}$)	9.94
Butylated hydroxytoluene ($C_{15}H_{24}O$)	23.01
Total	100

Extraction Investigations

The solvent extraction studies revealed that the extent of copper extracted for spent sludge liquor increases rapidly with increasing Cyanex®272 concentration and equilibrium pH at room temperature as follows:

(i) **Effect of equilibrium pH**: The effect of pH on the extraction of Cu from its aqueous solution by 0.2 mol/L Cyanex®272 in kerosene, experiments were performed in the equilibrium pH ranges of 1–5 at 27 ± 2 °C for 25 min. The results of the percent extraction of copper from 34.6 to 91.66% with increasing pH from 1 to 4 are summarized in Fig. 1.

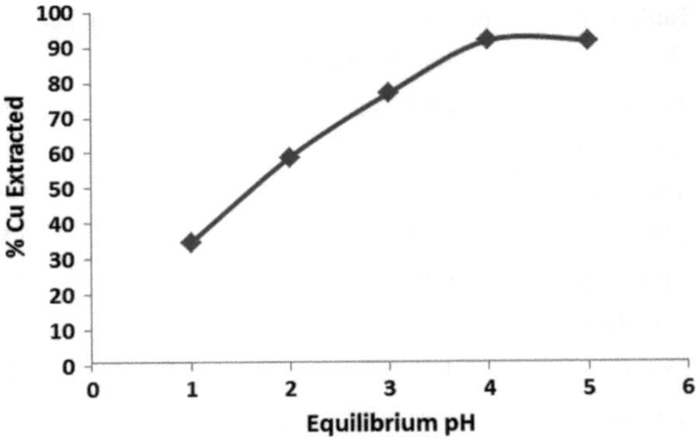

Fig. 1 Equilibrium pH variation on the percent of Cu extracted by 0.2 mol/L Cyanex®272 in kerosene

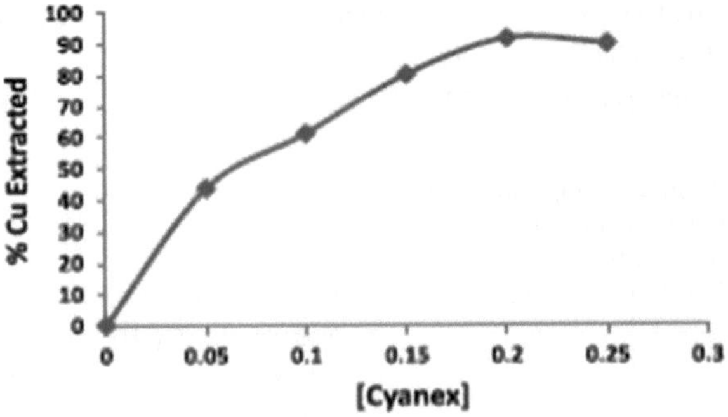

Fig. 2 Percent of Cu extracted by different Cyanex concentrations at pH 4.0

Increasing the equilibrium pH above 4 decreases the percent of Cu extracted to 91.50% at pH 5. Thus, the drastic reduction with increasing solution pH may be due to the fact that at high pH and apart from possible precipitation phenomena, there is a considerable increase in distribution coefficient indices [4, 16].

(ii) **Effect of Cyanex concentration**: The effect of Cyanex®272 concentrations on copper extraction in the processed DSTO at optimal leaching at pH 4.0 was examined. The percent of Copper extracted increases from 43.7 to 91.7% with increasing Cyanex concentration from 0.05 to 0.2 mol/L. Also, increasing the extractant concentration beyond 0.2 mol/L showed a decreasing extraction to 90.1% by 0.25 mol/L Cyanex solution (Fig. 2) [4].

Further purification of the copper solution at optimal extraction conditions was examined through combinations of precipitation and crystallization techniques to obtain a high-grade copper oxide (CuO: 96-900-9471; melting point = 1309 °C/1326 °C *industrial standard*) amenable for use as *p*-type semiconductor. The operational hydrometallurgical scheme summarizing the analytical techniques for the enrichment of spent deposited sludge for the recovery of pure copper solution beneficiated as CuO is shown in Fig. 3.

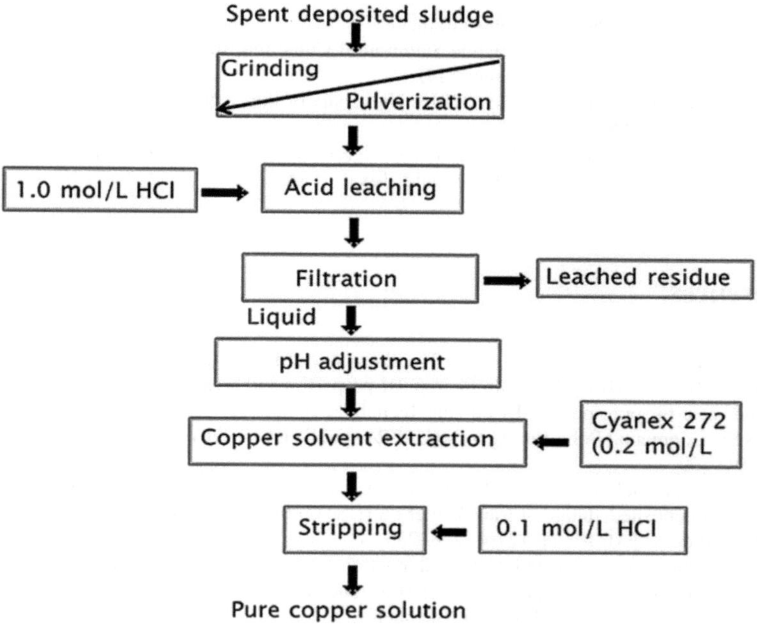

Fig. 3 A hydrometallurgical scheme for pure copper extraction from spent DSTO [4]

Conclusion

As the oil purification is required for keeping the transformer power voltage operating at optimal capacity, successful regeneration of pure oxidized transformer oils by combination of leaching and solvent extraction via hydrometallurgical route was examined in this study leading to the following conclusions:

(i) The detailed characterization of DSTO results gave variation of the transformer oil potency with respect to the ASTM D1819 regulations, indicating that the transformer oil is suitable for re-refining for industrial value addition;
(ii) The dissolution of heavy metals especially copper from spent DSTO by acid leaching and solvent extraction is possible;
(iii) At optimal leaching conditions, about 93% of initial 10 g/L DSTO sample was reacted within 120 min. Also, extraction of copper by Cyanex®272 increased with increasing extractant concentration and equilibrium pH. A 0.1 mol/L HCl solution was found to be efficiently stripped of about 94.8% copper from organic loaded DSTO containing copper [4];
(iv) The technique employed in this work proposed an efficient purification and regeneration of precious metals from spent DSTO, supporting recycling initiatives and employment creation.

Acknowledgements The authors wish to thank:
1. Dr. Oliver Rouher and Mrs. Christine Salomon of Cytec Industries, Rungis Cedex, France for their benevolence by supplying Cyanex®272;
2. The Central Research Laboratories (CRL), University of Ilorin, Nigeria for assisting in aqueous metal analyses by AAS; and
3. Tertiary Education Trust Fund (TETFund), Nigeria for sponsorship to attend the 2019 TMS Annual Meeting & Exhibition in San Antonio, Texas, USA.

References

1. Hutzinger O, Safe S, Zitco V (1974) The chemistry of PCBs. CRC Press, Cleveland, pp 7–39
2. Choi S, Huh C (2013) The lightning impulse properties and breakdown voltage of natural ester fluids near the pour point. J Electr Eng Technol 8(3):524–529
3. US Environmental Pollution Agency (USEPA) public law (1976). Toxic substance control act, pp 94–469
4. Baba AA, Ayodele JS, Ameen OM, Jimoh BA, Lawal A (2018) Characterisation and treatment of spent deposited sludge of transformer oil (DSTO) by acid leaching and solvent extraction. Miner Process Extr Metall 127(4):228–235
5. Liu D, Zhou Y, Yang, Zhang L (2016) Characterization of high performance AIN nano-particle-based transformer oil nanofluids. IEEE Trans Dielectr Electr Insul 23(5):2757–2767
6. Safiddine L, Zafour AH, Fofana I, Skender A, Guerbas F, Boucherit A (2017) Transformer oil reclamation by combining several strategies enhanced by the use of four adsorbents. IET Gener Transm Distrib 11:2912–2920
7. Gavrilous G, Borscevkis O (2011) Insulation oil treatment and its' necessity in power transformers, pp 23–27
8. Laurentino AC, Parize AL, Laranjeira MCM, Martins AR, Mayer NM, De Fa'vere VT (2007) Regeneration of insulating mineral oil by carbonated amorphous calcium phosphate-chitosan adsorbent. Process Saf Environ Prot 85(B4):327–331
9. Leclerc YN, Menx E, Wang YG (2003) Hydrometallurgical extraction of zinc ferrites. Hydrometallurgy 75:169–176
10. Turan MD, Hundogan HS, Tumen F (2004) Recovery of zinc and lead from zinc plant residue. Hydrometallurgy 75:169–176
11. Baba AA, Adekola FA (2011) Beneficiation of a Nigerian sphalerite mineral: solvent extraction of zinc by Cyanex®272 in hydrochloric acid. Hydrometallurgy 109:187–193
12. Somer G, Nakisci Unlu A (2006) The effect of acid digestion on the recoveries of trace elements: recommended policies for the elimination of losses. Turk J Chem 30:745–753
13. Baba AA, Ayinla IK, Adekola FA, Ghosh MK, Pradeep CR, Amos IA (2015) Extraction and purification of copper from a Nigerian chalcopyrite ore leach liquor by dithizone in kerosene. Solvent Extr Res Dev Japan 22(2):2135–2146
14. Ochromowicz K, Chmielewski T (2013) Solvent extraction of copper (II) from concentrated leach liquors. Physicochem Probl Miner Process 49(1):357–367
15. Baba AA, Adekola FA, Bale RB (2009) Development of a combined pyro—and hydrometallurgical route to treat zinc—carbon batteries. J Hazard Mater 181(3 4):837–844
16. Hosseini T, Mostoufi N, Danesphpayeh M, Rashichi F (2011) Modelling and optimization of synergistic effect of Cyanex 302 and D2EHPA on separation of zinc and manganese. Hydrometallurgy 105:227–283

Leaching and Recovery of an Oxide Gold Concentrate Using Ammoniacal Thiosulfate Solutions

Zhonglin Dong, Tao Jiang, Bin Xu, Yongbin Yang and Qian Li

Abstract Thiosulfate leaching-resin adsorption recovery, an environmentally friendly process, has been developed to efficiently extract gold from an oxide gold concentrate. 80.5% of gold extraction could be achieved by thiosulfate leaching under the conditions of $(NH_4)_2S_2O_3$ 0.2 mol/L, $CuSO_4$ 0.025 mol/L, NH_3 1.5 mol/L and leaching time 8 h, and the thiosulfate consumption was 53.6 kg/t-concentrate. IRA-400 anion exchange resin was used to recover gold from the leach solution, and an adsorption efficiency of 99.3% gold and 39.4% copper was attained. Two-step elution process, including pre-elution of copper with 0.5 mol/L $(NH_4)_2S_2O_3$ followed by gold elution with 2 mol/L NaCl + 0.5 mol/L Na_2SO_3 was adopted, and 98.6% of copper and 99.1% of gold could be successfully eluted from the resin.

Keywords Gold · Thiosulfate leaching · Recovery · Ion-exchange resin

Introduction

For the past century, cyanidation has been the predominant method for gold leaching from ores owing to its simple process and low cost [1]. However, there has been increasing public concern over the use of cyanide due to its strong toxicity. In addition, cyanide leaching shows unsatisfactory performance on refractory gold ores especially for those containing copper or "preg-robbing" carbon [2]. Thus, considerable attention has been paid to the alternative lixiviants, the most promising of which is thiosulfate. Compared with conventional cyanidation, thiosulfate leaching has the advantages of lower toxicity, lower reagent costs and faster leaching rate, particularly for some refractory ores, e.g. some carbonaceous gold ores [3].

Z. Dong · T. Jiang · B. Xu (✉) · Y. Yang · Q. Li
School of Minerals Processing and Bioengineering, Central South University, Changsha 410083, China
e-mail: xubincsu@csu.edu.cn

G. Azimi et al. (eds.), *Rare Metal Technology 2019*, The Minerals, Metals & Materials Series, https://doi.org/10.1007/978-3-030-05740-4_2

However, the successful commercial application of thiosulfate leaching is still rare, except for the development of an ammonia-free thiosulfate leaching process by Barrick Gold Corporation to treat a carbon-bearing sulfide gold ore pretreated with acidic or alkaline pressure oxidation [4]. Most of the past researches have concentrated on the gold leaching from its ores using thiosulfate solutions over the past several decades [5–11]. However, limited studies on gold recovery from the leach solutions have been conducted and there is a lack of a suitable recovery process. Past studies have pointed that it is difficult to recover gold from thiosulfate solutions using the common techniques of activated carbon adsorption, cementation, electrowinning, and solvent extraction. This can be attributed to the weak affinity of activated carbon for $[Au(S_2O_3)_2]^{3-}$ complex or the presence of large amount of undesirable anions particularly copper (I) thiosulfate complexes and sulfur-oxygen anions. In comparison, resin adsorption technique is more suitable because of its fast adsorption speed, high loading capacity, low requirements on the clarity of solutions, simultaneous elution and regeneration at ambient temperature through the elaborate choice of eluent. In addition, ion-exchange resins can be custom-made to selectively extract gold because the functional groups can be designed to have high affinity for objective ions in the solution [1, 2].

In this paper, a detailed study on the factors affecting the thiosulfate leaching of gold was carried out. Afterwards, the adsorption of gold and copper on the resin and their elution from the loaded resin were investigated. Based on these studies, the environmentally friendly process consisting of thiosulfate leaching-resin adsorption recovery was developed to efficiently extract gold from an oxide gold concentrate.

Experimental Work

Materials and Reagents

The original gold concentrate used in this study had the particle size of 95% less than 0.074 mm, and its chemical composition is indicated in Table 1. The gold content was 3.3 g/t, whilst the total content of acid gangues including SiO_2 and Al_2O_3 was 72.16%. The mineralogical phases of the concentrate were determined by quantitative X-ray diffraction analysis, and the result showed that the main minerals were quartz and muscovite. Table 2 shows the result of chemical phase analysis of gold. As indicated, 90.9% of the gold was exposed, and the others were mainly encapsulated in oxides and sulfides

Table 1 Chemical composition of the oxide gold concentrate (%)

Constituent	Au[a]	SiO_2	Al_2O_3	CaO	MgO	Fe	S	C	Ti	Mn	Cr
Content	3.3	58.20	13.96	6.86	4.72	5.25	0.90	0.75	0.90	0.07	0.06

[a]Unit g/t

Table 2 The chemical phases of gold in the oxide gold concentrate

Phases	Exposed gold	Encapsulated in oxides	Encapsulated in sulfides	Encapsulated in silicates	Total
Content g/t	3.0	0.2	0.08	0.02	3.3
Distribution %	90.9	6.1	2.4	0.6	100.00

The reagents used in this study, such as ammonium thiosulfate, copper sulfate, ammonia, sodium chloride, sodium sulfite, sodium hydroxide and sulfuric acid were of analytically grade. Ultrapure water was used throughout all experiments.

Leaching and Recovery Tests

All thiosulfate leaching tests were performed in 1 L baffled PVC reactor using an overhead stirrer with a flat-bladed impeller, which was open to air through a sampling port. For each leaching test, 0.5 L of distilled water containing desired quantities of requisite reagents was firstly transferred into the reactor, and then a given mass of the gold concentrate was added and the formed pulp was simultaneously agitated at a constant speed of 300 rpm. All of these tests were performed at a liquid-to-solid ratio of 3. Pulp pH was adjusted to a preset value with the addition of NaOH (1 mol/L solution) and was kept steady by readjustment every half an hour. Pulp temperature was kept at 25 ± 0.5 °C with a water bath during leaching. When the reaction was completed, pulp was vacuum filtrated. The obtained solution samples were immediately subjected to thiosulfate concentration assays and gold recovery tests, and the residue samples were analyzed for gold content to calculate the gold extraction.

Strong-base anion exchange resin Amberlite IRA-400 was adopted to recover gold from the leach solutions. First, a certain amount of resin (wet base value) was added into 100 mL of leach solution, which was agitated in a 250 mL beaker at 200 rpm. A 1 mL of the sample, if necessary, was consecutively taken to observe the concentration changes of gold and copper. When the adsorption was finished, the leach solution was filtrated by a vacuum filter and the obtained resin was adequately rinsed with ultrapure water. Then, the gold-loaded resin was transferred to an ion-exchange column which was constructed with a 10 mL burette with small wool plugs both on the top and at the bottom of the resin bed. Finally, the eluants were pumped into the column through a rubber stopper on the top of the burette using a peristaltic. The flow rate of eluants was 3BV/h (bed volume per hour) in this experiment where 1 BV was the volume of the burette occupied by the resin bed. During the elution stage, consecutive 1 mL of eluate was collected from the tip at the bottom of the burette for gold and copper assays. The elution efficiencies of copper and gold were calculated based on the content differences of these two metals on the loaded resin and in the eluate.

Analytical Methods

Sulfur and carbon contents in solid were determined using a high-frequency IR carbon and sulfur analyzer (HW2000B, Wuxi Yingzhicheng). The other elements in solid were analyzed using acid digestion and an atomic absorption spectrometer (AA-6800, SHIMADZU). Element concentrations in solution were all measured by an inductively coupled plasma-atomic emission spectrometer (PS-6, Baird). Thiosulfate concentration was determined by an iodometric method with the indicator Vitex, and certain amounts of EDTA-2Na and formaldehyde were added before the titration in order to eliminate the interferences of cupric tetra-amine complex and sulfite.

Mineralogical compositions of samples were obtained by an X-ray diffractometer (D/Max 2500, Rigaku). Chemical phase constitution analysis has been widely utilized in China to determine the distribution percentage of an element for its every phase in research sample, which is usually carried out by professionals in research institutes of mining and metallurgy. In this paper, the analysis procedure was as follows: At first, the phases of element were discovered using an optical microscope combined with an electron probe micro-analyzer. After that, selective dissolution was carried out to determine element contents in identified phases by appropriate process flow and chemical reagents.

Results and Discussion

Cupric-Ammonia Catalyzed Thiosulfate Leaching

Effect of Initial Thiosulfate Concentration

Figure 1 shows the effect of initial thiosulfate concentration on gold extraction and thiosulfate consumption. Only 44.2% of gold extraction was obtained when the thiosulfate concentration was 0.05 mol/L, the thiosulfate consumption was also only 21 kg/t-concentrate. Increasing the initial thiosulfate concentration from 0.05 to 0.15 mol/L evidently increased gold extraction as well as thiosulfate consumption. However, when the thiosulfate concentration increased to 0.25 mol/L, there was no evident increase in gold extraction, but thiosulfate consumption continued to increase to 52.3 kg/t-concentrate. Therefore, the optimum thiosulfate concentration is around 0.20 mol/L.

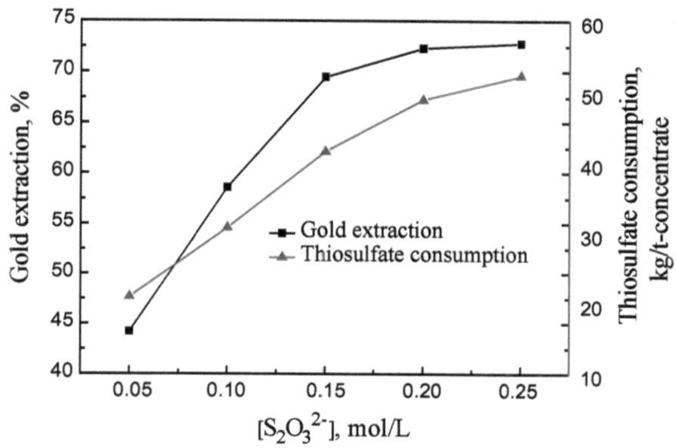

Fig. 1 Effect of initial thiosulfate concentration on gold extraction and thiosulfate consumption under the conditions: Cu^{2+} 0.02 M, NH_3 1.5 M and leaching time 8 h

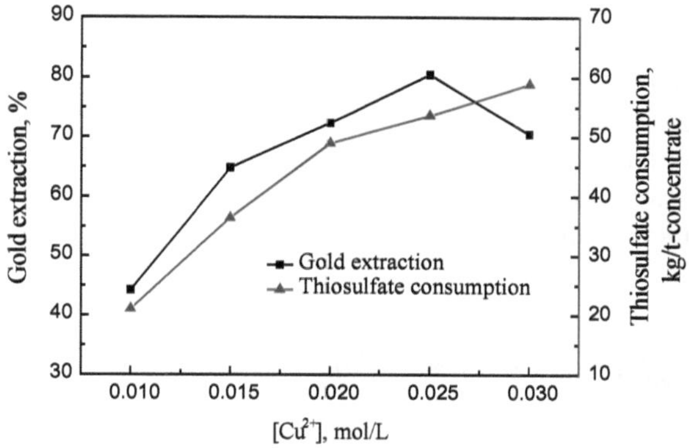

Fig. 2 Effect of initial cupric concentration on gold extraction and thiosulfate consumption under the conditions: $S_2O_3^{2-}$ 0.2 M, NH_3 1.5 M and leaching time 8 h

Effect of Initial Cupric Concentration

Figure 2 presents the effect of initial cupric concentration on gold extraction and thiosulfate consumption. Gold extraction increased with the increase of initial cupric concentration in the initial range of 0.01–0.025 mol/L. However, when the concentration was beyond the range, gold extraction began to decrease. The possible reason for this is that the high thiosulfate consumption led to the passivation of gold surface [1]. Thiosulfate consumption increased as the concentration increased in the entire abscissa range. Thus, the optimum cupric concentration is about 0.025 mol/L.

Effect of Initial Ammonia Concentration

Figure 3 shows the effect of initial ammonia concentration on gold extraction and thiosulfate consumption. Gold extraction increased with increasing ammonia concentration until it reached maximum at 1.5 mol/L and then decreased gradually until

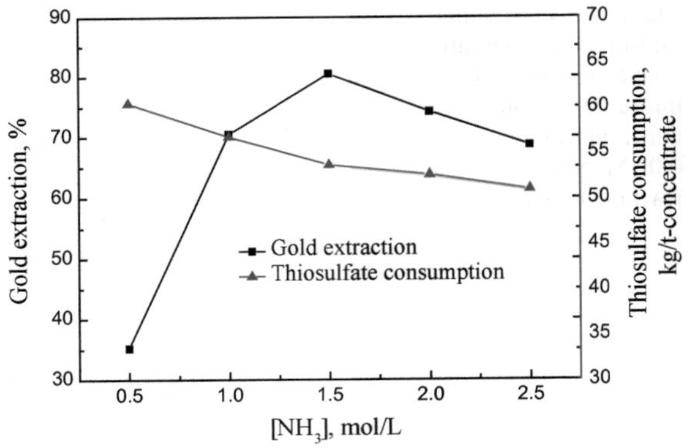

Fig. 3 Effect of initial ammonia concentration on gold extraction and thiosulfate consumption under the conditions: Cu^{2+} 0.025 M, $S_2O_3{}^{2-}$ 0.2 M and leaching time 8 h

2.5 mol/L. The evident decline of gold extraction can be attributed to the formation of $(NH_4)_5Cu(S_2O_3)_3$ under high ammonia concentration. This not only lowered the activity of $Cu(NH_3)_4^{2+}$ complex but also inhibited the gold leaching due to the cover of the solid precipitate on the gold particle surface [12] Thiosulfate consumption was always decreasing in the whole range of ammonia concentration studied. Therefore, the optimum ammonia concentration is around 1.5 mol/L.

Effect of Leaching Time

Figure 4 shows the effect of leaching time on gold extraction and thiosulfate consumption. Clearly, both gold extraction and thiosulfate consumption increased in the whole range of leaching time studied. From the above, the optimal leaching conditions were established as follows: $S_2O_3{}^{2-}$ 0.2 mol/L, Cu^{2+} 0.025 mol/L, NH_3 1.5 mol/L and leaching time 8 h. The gold extraction from the gold concentrate by thiosulfate leaching under these conditions was 80.5%, whilst the thiosulfate consumption reached up to 53.6 kg/t-concentrate.

Gold Recovery from the Leach Solution by the Ion-Exchange Resin

Adsorption of Gold and Copper on the Ion-Exchange Resin

The gold concentrate was first leached under the optimal conditions obtained from the above leaching experiment. Then, the pulp was filtrated, and the obtained solution sample was used for the following adsorption experiment. The concentrations of gold

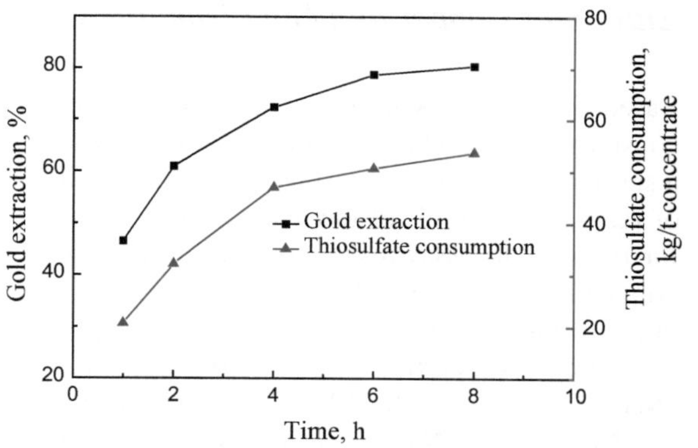

Fig. 4 Effect of leaching time on gold extraction and thiosulfate consumption under the conditions: Cu^{2+} 0.025 M, NH_3 1.5 M and $S_2O_3{}^{2-}$ 0.2 M

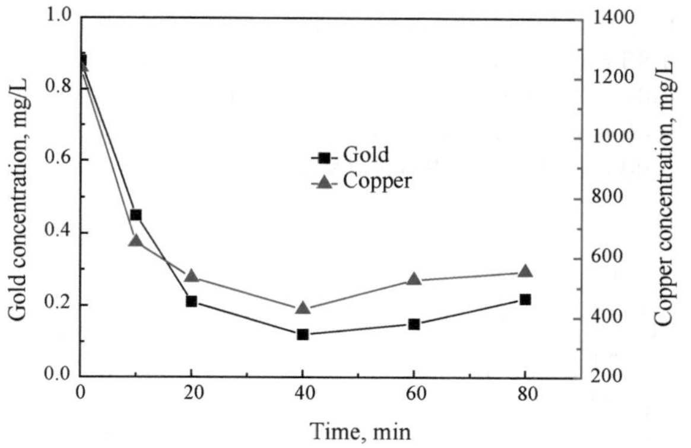

Fig. 5 Kinetic plots for the adsorption of gold and copper on IRA-400 anion exchange resin from the leach solution containing 0.88 mg/L gold and 1232.6 mg/L copper

and copper were separately 0.88 and 1232.6 mg/L after the solution was analysed. This indicates that parts of the copper precipitated and were transferred to the residue.

In this study, the adsorption behaviors of gold and copper on the resin were measured by adding 1 g of resin into 100 mL of the above leach solution, and the results are demonstrated in Fig. 5. As can be seen, the adsorption kinetics of gold and copper were fast, and both of their concentrations sharply decreased and simultaneously reached minimum at 40 min. It should be noticed that it's our goal to realize complete gold adsorption while minimizing copper adsorption. It can be seen, however, not all of the gold was adsorbed on the resin at this time. Furthermore, their concentrations increased again after a prolonged adsorption time, indicating that part of gold and copper were transferred from loaded resin into the leach solution. The potential reason is that high concentrations of polythionates mainly including trithionate ($S_3O_6^{2-}$) and tetrathionate ($S_4O_6^{2-}$) would be produced over an extended period due to the unwanted oxidation of thiosulfate in the leach solution. The two sulfur-oxygen anions could strongly adsorb on the resin and compete with the gold and copper thiosulfate complexes for ion-exchange sites [3]. Therefore, overlong adsorption time is not beneficial to the gold recovery.

Elution of Copper and Gold from the Gold-Loaded Resin

Based on the aforesaid adsorption tests of copper and gold on the resin, it was considered to use a larger amount of resin to realize the complete gold adsorption at a shorter adsorption time. Thus, the gold-loaded resin was prepared by adding 7 g of resin into 100 mL of the above leach solution, which was agitated at 200 rpm for 10 min. The assay results for the barren solution show that the concentrations of gold and copper were 0.006 and 747 mg/L, and therefore, the adsorption efficiency of them achieved 99.3 and 39.4%, respectively.

Elution was carried out in two stages: (1) pre-eluting copper with 0.5 M $(NH_4)_2S_2O_3$ solution for 12 BV, followed by (2) eluting gold with mixed solution of 2 M NaCl + 0.5 M Na_2SO_3 for 10 BV, and the results are shown in Fig. 6 and Table 3. It is clear that the elution profiles of copper and gold presented a sharp peak followed by a rapid decline, and the maximum concentration of them was separately 728 and 4.85 mg/L. As indicated in Table 3, the gold concentration in the copper eluate was only 0.001 mg/L, thus copper could be selectively eluted from the gold-loaded resin with negligible gold loss. The elution results show that the efficiencies of copper and gold could attain 98.6 and 99.1% after two-stage selective elution.

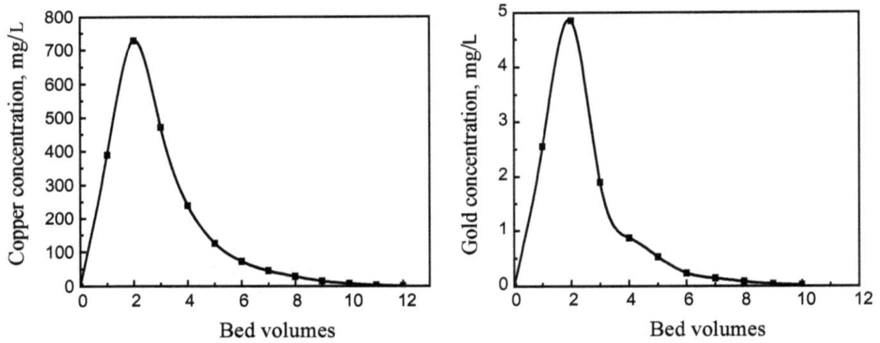

Fig. 6 Pre-elution of copper with 0.5 mol/L $(NH_4)_2S_2O_3$ followed by gold elution with 2 mol/L NaCl + 0.5 mol/L Na_2SO_3

Table 3 The elution results of copper and gold (the units of gold and copper concentrations are mg/L)

Copper eluate		Gold eluate		Copper elution efficiency, %	Gold elution efficiency, %
[Cu]	[Au]	[Au]	[Cu]	98.6	99.1
558.2	0.001	0.872	2.5		

Conclusions

The present work put forward a thiosulfate leaching-resin adsorption recovery process to extract gold from an oxide gold concentrate. 80.5% of gold in the concentrate was extracted by thiosulfate leaching under optimal conditions of $(NH_4)_2S_2O_3$ 0.2 mol/L, $CuSO_4$ 0.025 mol/L, NH_3 1.5 mol/L and time 8 h, whilst thiosulfate consumption was 53.6 kg/t-concentrate. An adsorption efficiency of 99.3% gold and 39.4% copper was achieved using 7 g of IRA-400 anion exchange resin to adsorb gold from 100 mL of the leach solution. Two-step elution process, consisting of pre-elution of copper with 0.5 mol/L $(NH_4)_2S_2O_3$ for 12 BV followed by gold elution with 2 mol/L NaCl + 0.5 mol/L Na_2SO_3 for 10 BV, was adopted to recover copper and gold from the loaded resin, and an elution efficiency of 98.6% of copper and 99.1% of gold could be attained.

Acknowledgements Financial supports from the National Natural Science Foundation of China (grant Nos. 51504293 and 51574284), the China Postdoctoral Science Foundation (grant No. 2014M550422), Hunan Provincial Natural Science Foundation of China (grant Nos. 2015JJ3149 and 2018JJ4038), and the Fundamental Research Funds for the Central Universities of Central South University (grant No. 502210012) are all gratefully acknowledged.

References

1. Xu B, Yang YB, Li Q, Jiang T, Liu SQ, Li GH (2016) The development of an environmentally friendly leaching process of a high C, As and Sb bearing sulfide gold concentrate. Miner Eng 89:138–147
2. Dong ZL, Jiang T, Xu B, Yang YB, Li Q (2017) Recovery of gold from pregnant thiosulfate solutions by the resin adsorption technique. Metals 7(12):555–571
3. Zhang H, Dreisinger DB (2002) The adsorption of gold and copper onto ion-exchange resins from ammoniacal thiosulfate solutions. Hydrometallurgy 66(1):67–76
4. Braul P (2013) Thiosulfate going commercial. Cim. Mag. 8:42–45
5. Breuer PL, Jeffrey MI (2002) An electrochemical study of gold leaching in thiosulfate solutions containing copper and ammonia. Hydrometallurgy 65(2):145–157
6. Arima H, Fujita T, Yen WT (2004) Using nickel as a catalyst in ammonium thiosulfate leaching for gold extraction. Mater Trans 45(2):516–526
7. Aylmore MG (2001) Treatment of a refractory gold-copper sulfide concentrate by copper ammoniacal thiosulfate leaching. Miner Eng 14(6):615–637
8. Chandra I, Jeffrey MI (2005) A fundamental study of ferric oxalate for dissolving gold in thiosulfate solutions. Hydrometallurgy 77(3):191–201
9. Senanayake G (2010) Gold leaching by copper (II) in ammoniacal thiosulfate solutions in the presence of additives. Part I: a review of the effect of hard–soft and Lewis acid–base properties and interactions of ions. Hydrometallurgy 96(4):1–20
10. Xu B, Yang YB, Li Q, Jiang T, Zhang X, Li GH (2017) Effect of common associated sulfide minerals on thiosulfate leaching of gold and the role of humic acid additive. Hydrometallurgy 17:44–52
11. Xu B, Yang YB, Jiang T, Li Q, Zhang X, Wang D (2015) Improved thiosulfate leaching of a refractory gold concentrate calcine with additives. Hydrometallurgy 152:214–222
12. Abbruzzese C, Fornari P, Massida R, Veglio F, Ubaldini S (1995) Thiosulphate leaching for gold hydrometallurgy. Hydrometallurgy 39(1–3):265–276

A Multi-step Process for the Cleaner Utilization of Vanadium-Bearing Converter Slag

Junyi Xiang, Guishang Pei, Qingyun Huang, Wei Lv, Mingrui Yang, Kai Hu and Xuewei Lv

Abstract A multi-step process has been developed for the recovery of metal values from vanadium-bearing converter slag. A beneficiation process combined with grinding, sieving and magnetic separation steps was firstly utilized to recover metallic iron from the vanadium-bearing converter slag. Then a calcification roasting–acid leaching process was developed to recover vanadium from the vanadium slag, forming vanadium-bearing solution and vanadium tailings. The vanadium-bearing solution can be further used to recover vanadium, and the vanadium tailings were treated by a carbothermic reduction–magnetic separation process, which generated two products, viz. ferroalloy and titanium concentrate. The ferroalloy can be used as raw materials for the steelmaking process whereas the titanium concentrate can be taken for Ti recovery. The results showed that the majority of iron, vanadium, chromium, and titanium in the vanadium-bearing converter slag can be fully recycled in such a comprehensive recovery process.

Keywords Vanadium slag · Beneficiation · Leaching · Carbothermic reduction Cleaner production

Introduction

China has the third largest vanadium reserves in the world while most of the vanadium exists as vanadium titano-magnetite ore [1]. Direct extraction of vanadium from vanadium titano-magnetite ore feedstock is not economical because of its low-grade quality. Generally, vanadium is first reduced into hot metal in a blast furnace, then oxidized in a converter furnace and enriched in a vanadium-bearing converter slag [2].

J. Xiang · G. Pei · W. Lv · M. Yang · K. Hu · X. Lv (✉)
School of Materials Science and Engineering, Chongqing Univesity, Chongqing 400044, China
e-mail: lvxuewei@163.com

Q. Huang
School of Metallurgical and Materials Engineering, Chongqing University of Science and Technology, Chongqing 401331, China

G. Azimi et al. (eds.), *Rare Metal Technology 2019*, The Minerals, Metals & Materials Series, https://doi.org/10.1007/978-3-030-05740-4_3

Vanadium-bearing converter slag contains a considerable amount of metal values, such as Fe, V, Cr, Ti, Mn, Ca, Mg and Al, that are useful for various applications [3]. The converter slag also contains a considerable amount of metallic iron, which might contribute to the sinter ring formation in the rotary kiln aroused from the local overheat problems during the roasting process [4]. Therefore, the metallic iron should be removed before the subsequent extraction process. It will also dissolve in the sulfuric solution and increase the burden of purification. Furthermore, the recycling of metallic iron can also improve the iron and steelmaking process due to the content of metal values in the recycled materials [5].

The most mature and commercial vanadium extraction technology for vanadium-bearing converter slag is the sodium salt roasting–water leaching process [6]. Unfortunately, this technology leads to a series of problems, such as fusion agglomeration of the slag at high roasting temperatures, emission of corrosive gases (HCl, Cl_2, SO_2 and SO_3), discharge of huge quantities of sodium-bearing wastewater, and utilization difficulty of the tailings [7]. Calcification roasting–acid leaching is an alternative process with less gas emission, less harmful wastewater and little noxious solid waste poll [8].

The majority of the vanadium are recovered from the vanadium extraction process, almost all of the iron, chromium, titanium, and other elements remain in the vanadium tailings. The enormous quantity of vanadium tailings discarded annually not only occupies large areas of land, but also accompanies with high disposal costs [9]. Vanadium tailings can be used to produce geopolymers [10], V-Ti black ceramic [11], far-infrared radiation coatings [12], but only very small part of tailings was used every year. The recovery of valuable metals from vanadium tailings can not only increase the value of the tailings, but also minimize the amount of hazardous elements. In this study, a multi-step process was utilized for the recovery of metal values from the vanadium-bearing converter slag for the purpose of comprehensive utilization and cleaner production.

Experimental

Experimental Materials

The vanadium-bearing converter slag used in the present study is obtained from China Panzhihua Iron and Steel Group Corp. The backscattered electron (BSE) image of the polished section of the received vanadium-bearing converter slag is shown in Fig. 1. It can be seen that the received converter slag mainly consists of three distinct mineral phases, viz. metallic iron, spinel, and olivine, which could be confirmed by the EDS measurements. The bulk chemical composition of the converter slag used in this study is reported in Table 1. All the other reagents used for leaching, reduction and chemical analysis were of analytical grade and of purity greater than 99.9%.

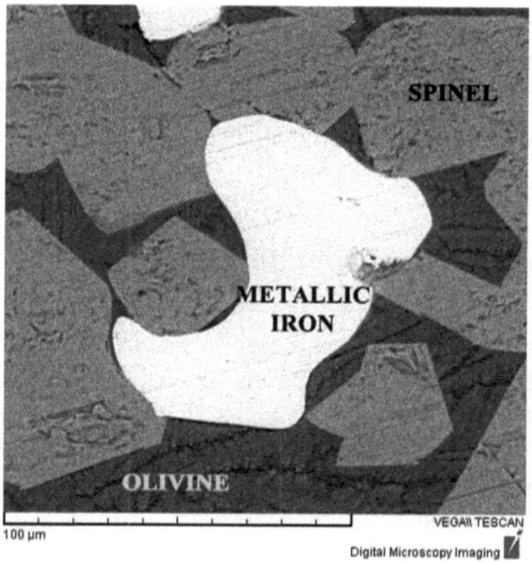

Fig. 1 Backscattered electron (BSE) image of the received vanadium-bearing converter slag

Table 1 Chemical composition of the received vanadium-bearing converter slag (wt%)

Fe	Ti	V	Cr	Si	Mn	Ca	Mg	Al
42.10	6.96	6.96	1.40	5.59	4.60	0.44	0.41	0.42

Samples Preparation

The received converter slag was firstly crushed by a jaw crusher (EP-1, Henan, China) and vibration mill (3MZ-100, Jiangxi, China) for the liberation of metallic from the slag phases and then sieved by a series of sieves with different aperture sizes. After that, the magnetic particles were separated from the sieved samples by using a low-intensity magnetic separator with the magnetic field intensity of 2000 gauss.

Vanadium Extraction

The separated vanadium slag was firstly mixed with a determined amount of calcium carbonate and then roasted in an open muffle furnace at 850 °C for 2 h in air. The mole ratio of vanadium to calcium in the mixture was maintained at 1:1. After roasting, the samples were removed from the furnace and cooled to room temperature in air. The roasted samples were milled to a particle size smaller than 75 μm and then leached in a sulphuric acid solution with a constant pH value of 2.5, 50 °C leaching temperature, 150 rpm stirring speed, and 1:10 solid to liquid ratio. A three-necked round-bottomed flask equipped with a water condenser, thermometer, pH meter, thermostatic water bath, and magnetic stirrer was used for leaching. Sulfuric acid with a concentration of 15 pct was used to maintain the acidity of the leaching solution throughout the leaching process. After a specific leaching time, the slurry was filtered and the leaching residue was dried at 100 °C for 24 h.

Carbothermic Reduction–Magnetic Separation

The mixture of the leaching residue and carbon was compressed into cylinders with both diameter and height of approximately 20 mm. The cylindrical sample was initially dried at 120 °C for 24 h, then roasted in a vertical-tube electric furnace at 1400 °C for 1 h. A flow rate of 500 ml/min argon (with 99.999% purity) was maintained through the furnace during the entire heating process. After roasting, the reduced cylindrical sample was removed from the furnace and cooled to ambient temperature under an argon atmosphere, then crushed to a particle size smaller than 125 μm. Then a Davis Magnetic Tube (XCGS-50, Shicheng, China) was used to separate the magnetic portion from the crushed powders with a magnetic field intensity of 120 kA/m.

Analysis

The chemical compositions of the samples were established with the help of inductively coupled plasma-optical emission spectrometry (ICP-OES), using an Optima 8000 Instrument (PerkinElmer, USA). The scanning electron microscopy (SEM) and energy-dispersive X-ray spectroscopy (EDS) analysis of the samples was performed using a scanning electron microscope (VEGA 3 LMH; TESCAN) equipped with an energy-dispersive X-ray spectroscope (Oxford).

Results and Discussion

Recovery of Metallic Iron from the Converter Slag

In order to liberate the metallic iron from the converter slag, the received vanadium-bearing converter slag should be thoroughly ground. Figure 2 shows the particle size distribution, magnetic separation yields, and vanadium contents for the ground and sieved samples.

As shown in Fig. 2, the proportion of the magnetic particles is significantly increased with the increase of the particle size. Nearly, all of the fines (< 0.075 mm) are non-magnetic, while all of the oversized particles (> 4 mm) are magnetic. It's

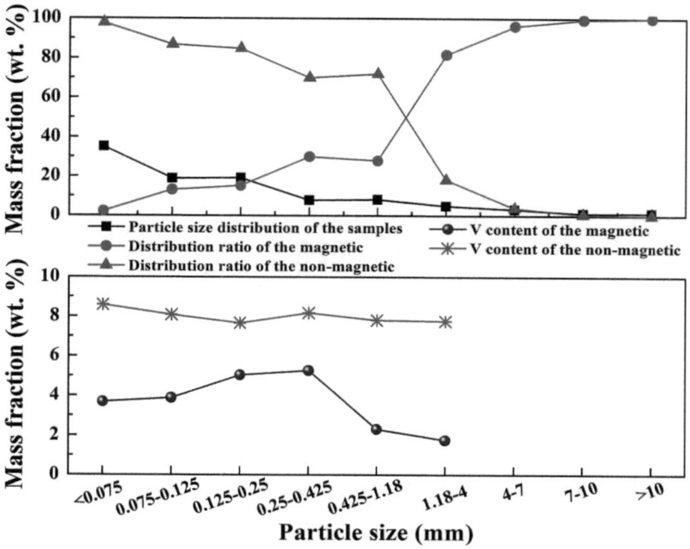

Fig. 2 The particle size distribution, magnetic separation yields, and vanadium contents for the ground and sieved samples

obvious that the vanadium content increased from 3.69 to 5.03% then decreased to 1.75% with the increasing particle size of the magnetic. This indicates that the considerable amount of metallic iron still didn't liberate from the classified converter slag with the particle size of 0.125–4 mm. Therefore, this part should be crushed and separated again to increase the recovery of vanadium. The metallic iron was thoroughly liberated from the classified converter slag with the particle size of 0.125–4 mm by secondary grinding, sieving and magnetic separation. All of the separated non-magnetic parts were well mixed as the raw materials for the vanadium extraction process. The chemical composition of the mixed non-magnetic part ("vanadium slag" is used hereafter) is shown in Table 2, the content of total iron significantly decreased from 42.1% to about 31% and the content of vanadium increased from 6.96% to about 8.57% after the separation process.

Recovery of Vanadium from the Vanadium Slag

After the removal of metallic iron from the converter slag, vanadium was enriched in the vanadium slag, which can be recovered by the calcification roasting–acid leaching process. Vanadium slag is roasted with limestone to transform vanadium-bearing spinels into calcium vanadates and followed by leaching with acid as the following reactions [13]:

Table 2 Chemical composition of the obtained solid samples

Compound	Chemical composition (wt%)								
	Fe	Ti	V	Cr	Si	Mn	Ca	Mg	Al
Vanadium slag	31.00	8.63	8.57	0.98	6.77	5.67	1.84	1.75	1.84
Roasted slag	27.35	7.64	7.40	0.86	5.99	4.95	6.11	1.53	1.59
Vanadium tailings	31.80	8.93	1.49	1.05	6.77	3.78	4.81	1.54	1.78
Ferroalloy	70.78	8.05	3.00	2.20	/	/	/	/	/
Ti concentrate	0.92	16.53	0.44	0.20	/	/	/	/	/

$$4FeV_2O_4 + 5O_2 + 8CaCo_3 \rightarrow 4Ca_2V_2O_7 + 8CO_2 + 2Fe_2O_3 \quad (1)$$

$$Ca_2V_2O_7 + 2H_2SO_4 \overset{pH = 2.5}{\longrightarrow} H_4V_2O_7 + 2CaSO_4 \quad (2)$$

As shown in Table 2, the content of vanadium slightly decreased from 8.57 to 7.40% after the calcification roasting process. The main phases in the roasted slag would transform from spinel and olivine to hematite, pseudobrookite, and calcium vanadate [14].

Figure 3 shows the variation of vanadium leaching rate with the leaching time. It's obvious that the leaching rate of vanadium is significantly influenced by the leaching time. The leaching rate of vanadium dramatically increased to about 70% in the initial 10 min, then gradually increased to about 83% with the extending of leaching time to 60 min. The consumption of the sulfuric acid (15%), used to maintain the acidity, shows that nearly all of the sulfuric acid are consumed in the first 20 min. This is consistent with the leaching behavior of vanadium. After leaching and filtrating, the leaching solution can be used to extract vanadium pentoxide by a vanadium recovery processes such as ammonium salt [7], solvent extraction [15, 16] or ion exchange [17]. The vanadium content in the leaching residue ("vanadium tailings" is used hereafter) decreased to 1.49% after the leaching process, whereas other compositions nearly insoluble (as shown in Table 2).

Recovery of Ferroalloy from the Leaching Residue

Recovery of metal values from the vanadium tailings will not only improve the value of tailings but also minimize the content of hazardous elements. Carbothermic reduction–magnetic separation process is an efficient way to recover iron, chromium and vanadium from the vanadium tailings. It is well established that vanadium and chromium are very strong carbide-forming elements [18]. Thus, the recovery of

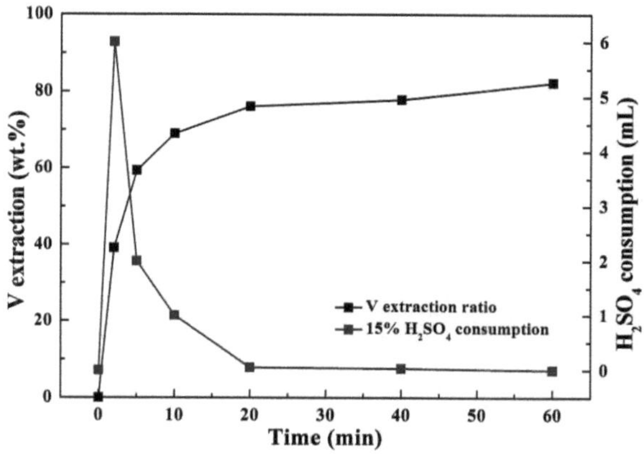

Fig. 3 The variation of vanadium leaching rate with time

vanadium and chromium is influenced by the carbon dosage during the reduction and magnetic separation process (as shown in Fig. 4). As the carbon ratio was increased from 8 to 11%, a gradual increase was observed in the recovery of the iron, from 80.72 to 98.70% then remained relatively constant. Both of vanadium and chromium recoveries significantly increased from less than 20% to approximately 90% with increasing carbon ratio from 8 to 12% then remained constant. However, the yield of magnetic part gradually increased from 32.6 to 61.5% with the increase of carbon ratio from 8 to 13%. High proportion of carbon in charge is favorable to the recovery of vanadium and chromium as alloying elements of metallic iron. However, excessively high carbon ratio will hinder the liberation of metallic iron and leads to the yield of magnetic part far more than the theoretical value.

Following the carbothermic reduction and magnetic separation processes, almost all of the iron, approximately 90% of the vanadium, and 94.6% of the chromium can be recovered by ferroalloy which can be used as raw materials of the steelmaking process. As shown in Table 2, the non-magnetic portion (Ti concentrate) is enriched with titanium and possesses trace ions; therefore, it is considered a valuable raw material for extracting titanium via hydrometallurgy.

Evaluation

Based on these studies, a process flow sheet has been developed to process vanadium-bearing converter slag for fully recovering valuable metals and is shown in Fig. 5. In this process, 41.5% of iron can be recovered in the beneficiation step, and approximately 58% can be recovered in the carbothermic reduction-magnetic separation step. 70.21% of vanadium can be recovered as the vanadium-bearing solution, and 12.95% can be recycled as the ferroalloy. 57.2% of chromium are remained in the

Fig. 4 Recoveries of metals and yield of magnetic part as a function of carbon ratio

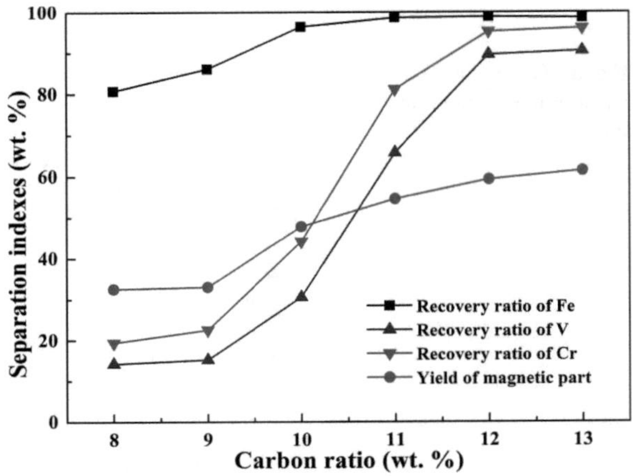

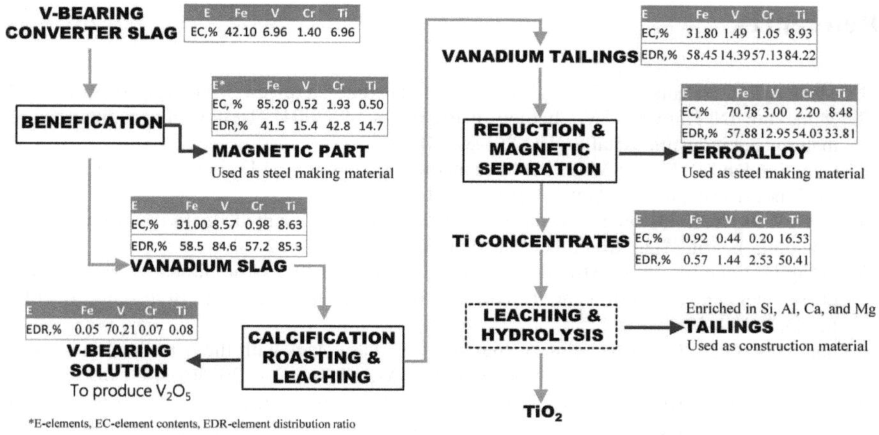

Fig. 5 Flow sheet of the multi-step process for the utilization of vanadium-bearing converter slag

vanadium slag through the beneficiation step, but then nearly all of them can be recovered as the ferroalloy. Approximately 50.5% of titanium can be recycled as the titanium concentrate with a content of 16.53% Ti.

Conclusions

A multi-step process was presented for the comprehensive utilization of vanadium-bearing converter slag. In this process, the metallic iron was firstly removed from the converter slag by a series of grinding, sieving and magnetic separation. After beneficiation, the content of iron in the vanadium slag significantly decreased from 42.2 to 31.0%, while the content of vanadium increased from 6.96 to 8.57%. Then a calcification roasting–acid leaching process was developed to recover vanadium from the vanadium slag, forming vanadium-bearing solution and vanadium tailings. The vanadium-bearing solution was further used to recover the vanadium, and the vanadium tailings was treated by a carbothermic reduction–magnetic separation process, which generated two products, viz. ferroalloy and Ti concentrate. Almost all of the iron, approximately 90% of the vanadium and 94.6% of the chromium remained in the vanadium tailings can be recovered by ferroalloy which can be used as raw materials in the steelmaking process. Whereas titanium was enriched in the titanium concentrate with a titanium content of 16.53%, which can be considered a valuable raw material for extracting titanium via hydrometallurgy.

Acknowledgements This work was supported by the project funded by China Postdoctoral Science Foundation and the Natural Science Foundation of China [grant number 51404047].

References

1. Moskalyk R, Alfantazi A (2003) Processing of vanadium: a review. Miner Eng 16(9):793–805
2. Monakhov IN, Khromov SV, Chernousov PI, Yusfin YS (2004) The flow of vanadium-bearing materials in industry. Metallurgist 48(7–8):381–385
3. Chand S, Paul B, Kumar M (2016) Sustainable approaches for LD slag waste management in steel industries: a review. Metallurgist 60(1):116–128
4. Li D, Wang X, Li Q, Zeng Z, Wang J (2000) Separation of magnetic materials from converter vanadium-bearing slag. Iron Steel Vanadium Titanium 21(2):52–54
5. Wang D, Jiang M, Liu C, Min Y, Cui Y, Liu J, Zhang Y (2012) Enrichment of Fe-containing phases and recovery of iron and its oxides by magnetic separation from BOF slags. Steel Res Int 83(2):189–196
6. Sadykhov GB (2009) Oxidation of titanium-vanadium slags with the participation of Na_2O and its effect on the behavior of vanadium. Russ Metall (Metally) 2008(6):449–458
7. Li HY, Wang K, Hua WH, Yang Z, Zhou W, Xie B (2016) Selective leaching of vanadium in calcification-roasted vanadium slag by ammonium carbonate. Hydrometallurgy 160:18–25
8. Liu J, Li L, Zheng S, Wang S, Du H, Xie H (2014) Extraction of vanadium from vanadium-containing slag by roasting-hydrothermal alkali leaching. Chin J Process Eng 14(05):763–769
9. Aarabi-Karasgani M, Rashchi F, Mostoufi N, Vahidi E (2010) Leaching of vanadium from LD converter slag using sulfuric acid. Hydrometallurgy 102(1–4):14–21
10. Wei B, Zhang Y, Bao S (2017) Preparation of geopolymers from vanadium tailings by mechanical activation. Constr Build Mater 145:236–242
11. Yang YG, Xu JH, Cai B, Wang QC, Xiu DP, Zhao ZB, Sun QZ, Cao SL (2013) Synthesis and applications of black ceramic from recycled industrial wastes. Adv Appl Ceram 112(3):146–148
12. Hao J, Liu A, Ma M (2009) Research of vanadium tiling's based far-infrared radiation coatings. Paint Coatings Ind 09:13–15
13. Kozlov VA, Demidov AE (2000) Chemical principles of a technology for making pure vanadium pentoxide. Metallurgist 44(8):428–433
14. Xiang JY, Huang QY, Lv XW, Bai CG (2017) Effect of mechanical activation treatment on the recovery of vanadium from converter slag. Metall Mater Trans B 48(5):2759–2767
15. Ma Y, Wang X, Wang M, Jiang C, Xiang X, Zhang X (2015) Separation of V(IV) and Fe(III) from the acid leach solution of stone coal by D2EHPA/TBP. Hydrometallurgy 153:38–45
16. Alibrahim M, Shlewit H, Alike S (2008) Solvent extraction of vanadium (IV) with di(2-ethylhexyl) phosphoric acid and tributyl phosphate. Period Polytech-Chem 52(1):29–33
17. Zipperian DC, Raghavan S (1985) The recovery of vanadium from dilute acid sulfate solutions by resin ion exchange. Hydrometallurgy 13(3):265–281
18. Bratberg J, Frisk K (2004) An experimental and theoretical analysis of the phase equilibria in the Fe-Cr-V-C system. Metall Mater Trans A 35(12):3649–3663

Efficient Extraction of V(V) in Aqueous Solution by Microemulsion System

Yun Guo, Danqing Li, Bing Xie and Hong-Yi Li

Abstract Vanadium slag is the main raw material to produce vanadium product. The mainstream approach that has been used to extract vanadium from vanadium slag is roasting–leaching. The above leaching solution through precipitation of ammonium polyvanadate (APV), APV-precipitated wastewater was produced. APV-precipitated wastewater contains a certain concentration of vanadium, which needs to require further treatment. In this research, we put forward an efficient method for the extraction of V(V) in aqueous solution using an Aliquat 336/isoamyl alcohol/n-heptane/NaCl microemulsion system. Suitable conditions of extracting vanadium in aqueous solution were studied, including surfactant concentration, the volume ratio of aqueous phase to microemulsion, as well as contacting time. And the extraction mechanism on the microemulsion system has been researched. Extraction of V(V) in APV-precipitated wastewater by microemulsion system also has been investigated; the extraction ratio of V can reach to 99.74%.

Keywords Microemulsion · Extraction · Vanadium · APV-precipitated wastewater

As an important rare metal, vanadium has been increasingly applied on many fields [1]. At present, vanadium slag is the main raw material to produce vanadium product [2]. The general process of extracting vanadium from vanadium slag is the following: vanadium slag roasted with sodium or calcium salts, leached with water, acidic or alkaline solution, precipitation of ammonium polyvanadate (APV) and ignition to obtain vanadium pentoxide [3–6]. After precipitation of ammonium polyvanadate, APV-precipitated wastewater was produced. APV-precipitated wastewater has serious toxicity due to the presence of vanadium. That vanadium extracted from APV-precipitated wastewater both can use resources fully and solve the environ-

Y. Guo · D. Li · B. Xie · H.-Y. Li (✉)
College of Materials Science and Engineering, Chongqing University, Chongqing 400044, China
e-mail: hongyi.li@cqu.edu.cn

Y. Guo · D. Li · B. Xie · H.-Y. Li
Chongqing Key Laboratory of Vanadium-Titanium Metallurgy and New Materials, Chongqing 400044, China

G. Azimi et al. (eds.), *Rare Metal Technology 2019*, The Minerals, Metals & Materials Series, https://doi.org/10.1007/978-3-030-05740-4_4

mental problem. So it is of importance to do research on how to extract vanadium from aqueous solution more efficiently.

The process of extracting vanadium from aqueous solution by microemulsion was studied in this paper. Microemulsion is thermodynamically stable, macroscopically homogeneous mixtures of at least three components, which consisting of water, organic solvent, and surfactant, occasionally with alcohol as a cosurfactant [7]. There are four types of microemulsion systems called Winsor I, Winsor II, Winsor III, and Winsor IV. Among these systems, Winsor I microemulsion, corresponding to oil in water phase coexisting with an excess oil phase, Winsor II microemulsion, corresponding to water in oil phase coexisting with an excess water phase, Winsor III, corresponding to both oil and water are interwoven, and a single-phase microemulsion (Winsor IV) [8]. Microemulsion has been widely applied in the filed of extraction of metal ions, such as gold [9, 10], rhenium [11], palladium [12], cobalt [13] and other metals. The above examples demonstrate the feasibility and effectiveness of using microemulsion for the extraction of metal ions. Winsor II system often effectively accelerates the extraction because of the enormous rise in the micro-interfacial surface area and the spontaneous formation of microemulsion globules. The globules can transport metal ions from the aqueous phase to the organic phase [14].

The goal of this article is to extract V from an acid solution by microemulsion system and provide the useful theory and experimental data. Firstly, we utilized the vanadium-containing simulated solution to explore the extraction mechanism and the extraction conditions, and then took APV-precipitated wastewater as an application object. There is much relation between the change in mass-charge ratio (N/Z) of vanadate ions in different existing forms and pH of aqueous solution. High N/Z ratio of vanadate ions leads to high extraction ratio of V since the extraction lies in electrostatic attraction force between vanadate ions and extractant on liquid–liquid interface. And because the APV-precipitated wastewater has strong acidity, we choose the pH of the synthetic solution at 3. It can be seen from Fig. 1 [15] that the V(V) in aqueous solution (pH = 3) is mostly in the forms $H_2V_{10}O_{28}{}^{4-}$. In this paper, Aliquat 336 played double functions of cationic surfactant and extractant. The influence of different factors on the extraction ratio (E%) was investigated, including surfactant concentration, the volume ratio of aqueous phase to microemulsion as well as contacting time. The results show that the application of microemulsion to extract V from aqueous solution is an extremely effective method.

Experimental

Instruments and Reagents

Inductively Coupled Plasma-Atomic Emission Spectrometry (ICP-AES, Optima 4300 DV, Perkin Elmer, USA), FE20 FiveEasy Plus pH meter (Switzerland), Electronic Balance (FA2004N, Mettler Toledo), Constant Temperature Heating Mag-

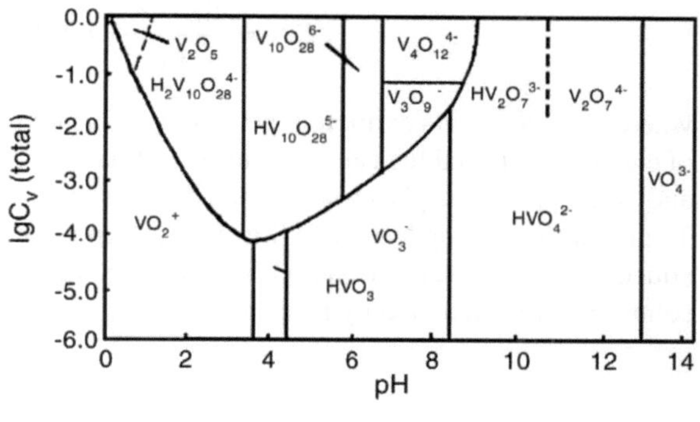

Fig. 1 Forms of vanadium (V) existing in aqueous solution

netic Stirrer (DF-101S, Gong Yi Electronic Equipment Limited Liability Company), Aquelix5 Pure Water Meter (ZTLC00031, Millipore Corporation).

Sodium orthovanadate dodecahydrate (CP, Sinopharm Chemical Reagent Co., Ltd), sodium hydroxide (AR, Chengdu Kelong Chemical Reagent Factory), sodium chloride (AR, Chengdu Kelong Chemical Reagent Factory), Sulfuric acid (AR, Chongqing Chuandong Chemical Co., Ltd), isoamyl alcohol (AR, Chengdu Kelong Chemical Reagent Factory), tricaprylyl methyl ammonium chloride (Aliquat 336; AR, Zhengzhou Hecheng New Material Technology Co., Ltd.) and n-heptane (AR, Chengdu Kelong Chemical Reagent Factory).

All aqueous solutions were prepared using distilled water.

Preparation of Microemulsion and Feed Solutions

The organic phase was prepared by injecting Aliquat 336, 0.5 mol/L sodium chloride and 0.5 mol/L sodium hydroxide solutions, isoamyl alcohol in n-heptane. The resulting mixtures form microemulsion spontaneously whenever the composition is adequate.

The synthetic solution was prepared by adding $Na_3VO_4{\cdot}12H_2O$ to deionized water. The pH of the aqueous solution was adjusted by H_2SO_4 and NaOH. In the following studies, the initial vanadium concentration is 1 g/L.

Experimental Methods

In the following studies, unless stated specially, the temperature was generally maintained at 298 $\pm$ 1 K, isoamyl alcohol (1.362 mol/L), Aliquat 336 (0.112 mol/L), the pH of aqueous solution is 3, and R = 4. The phase volume ratio was expressed as R. The extraction ratio ($E\%$) can be calculated according to the following equation:

$$E\% = \frac{C_0 - C_t}{C_0} \times 100\% \tag{1}$$

where $E\%$ represents extraction percentage, C_0 and C_1 represent the concentrations of elements in initial feed and extraction raffinate, respectively. The aqueous solution and microemulsion phase were mixed and shaken in a beaker using a Constant Temperature Heating Magnetic Stirrer at 200 rpm for 5 min. The V concentrations in aqueous solution were measured by ICP-AES, and the V concentrations in organic solution were calculated by mass balance.

Results and Discussion

The Effect of Aliquat 336 Concentration on the Extraction

The effect of concentration of Aliquat 336 in the microemulsion system on the extraction of V was depicted in Fig. 2. The Aliquat 336 concentration is 0.018 mol/L, when the extraction ratio of V is at 96.47%. The extraction ratio of V keeps on rising as the Aliquat 336 density was in the range of 0.018–0.073 mol/L. This was due to the amount of microemulsion produced increased with increasing surfactant and extractant (Aliquat 336) concentrations which could accelerate the extraction and improve the extractability of vanadium. With the Aliquat 336 concentration from 0.073 to 0.112, the extraction ratio of V remained approximately constant. At the certain range of the Aliquat 336 concentration, the extraction ratio of V can reach to 99.90%.

The extracted V(V) by Aliquat 336/isoamyl alcohol/n-heptane/NaCl microemulsion system was also studied. By plotting the logarithm of the V(V) distribution ratio

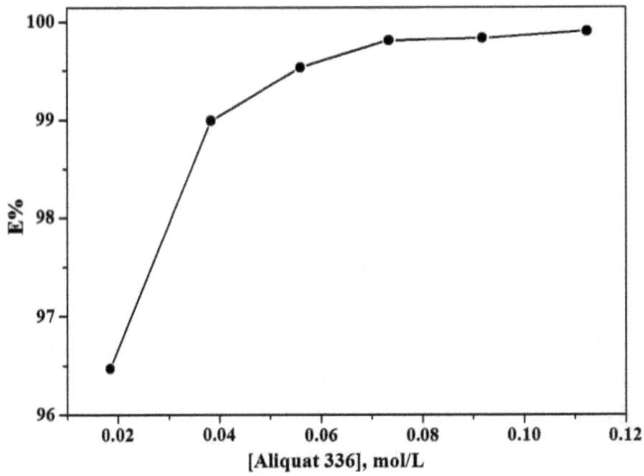

Fig. 2 Effect of concentration of Aliquat 336 on the extraction ratio of V

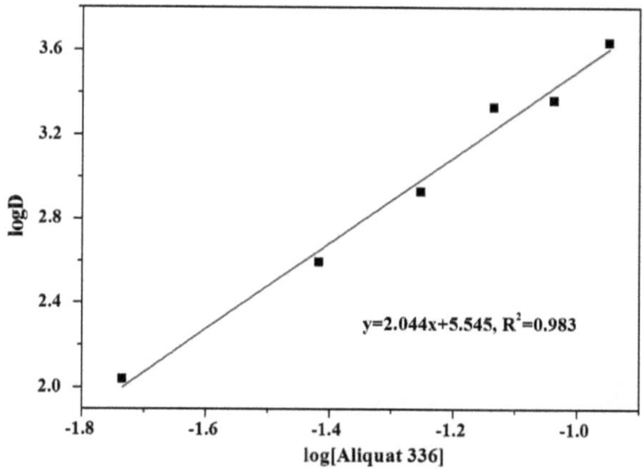

Fig. 3 Natural logarithm of distribution ratio (D) versus the natural logarithm of Aliquat 336 concentrations

(D) as a function of the logarithm of Aliquat 336 concentration, a straight line was obtained, as shown in Fig. 3.

The extraction reaction of V(V) by Aliquat 336/isoamyl alcohol/n-heptane/NaCl microemulsion system is presented as follows:

$$n\left[R_3CH_3N^+\right]\left[Cl^-\right]_{(o)} + H_2V_{10}O^{4-}_{28(a)} + (4-n)H^+_{(a)} = \left[R_3CH_3N^+\right]_n\left[H^+\right]_{4-n}\left[H_2V_{10}O^{4-}_{28}\right]_{(o)} + nCl^-_{(a)} \quad (2)$$

The equilibrium constant is:

$$K_e = \frac{([R_3CH_3N^+]_n[H^+]_{4-n}[H_2V_{10}O^{4-}_{28}]) \times [Cl^-]^n}{([R_3CH_3N^+][Cl^-])^n([H_2V_{10}O^{4-}_{28}])[H^+]^{(4-n)}} \quad (3)$$

and the distribution ratio is:

$$D = \frac{([R_3CH_3N^+]_n[H^+]_{4-n}[H_2V_{10}O^{4-}_{28}])}{([H_2V_{10}O^{4-}_{28}])} \quad (4)$$

$$K_e = \frac{D \times [Cl^-]^n}{([R_3CH_3N^+][Cl^-])^n[H^+]^{(4-n)}} \quad (5)$$

$$\log(D) = \log K_e + n\log\left(\left[R_3CH_3N^+\right]\left[Cl^-\right]\right) + (4-n)\log\left[H^+\right] - n\log\left[Cl^-\right] \quad (6)$$

It can be seen that the slope of log$D$ versus log[Aliquat 336] is 2.044 in Fig. 3, which is close to 2. It was indicated that vanadate coordinated with two extractant molecules (Aliquat 336) in complex. Decavanadate anions with various degrees of protonation were crystallized with tetraalkylammonium cations; a tetraprotonated polymer $H_4V_{10}O^{2-}_{28}$ assembly patterns have been observed [16]. Therefore and based on slope analysis of the experimental results, the extraction of V(V) from aqueous solutions was accepted as the formation of the complex as $[R_3CH_3N^+]_2[H_4V_{10}O_{28}^{2-}]$. The extraction reaction of V(V) by microemulsion can be written as:

$$2[R_3CH_3N^+][Cl^-]_{(o)}+H_2V_{10}O_{28(a)}^{4-}+2H_{(a)}^+ = [R_3CH_3N^+]_2[H_4V_{10}O_{28}^{2-}]_{(o)}+2Cl_{(a)}^- \quad (7)$$

From above, we can conclude that the extraction mechanism of V with microemulsion was proved to be an anion exchange reaction.

The Effect of Contacting Time on the Extraction

From former studies, we may know that the extraction process can be completed in 5 min by observing the color of the solution. In order to establish the optimum time required to accomplish equilibrium, the effect of contacting time on the extraction ratios of V was investigated in Fig. 4. The extraction ratio of V reaches 98.72% within 1 min of contacting time. The extraction ratio of V reaches peak value after 4 min of contacting time. It was found that the extraction equilibrium is immediately achieved.

The Effect of R on the Extraction

The effect of R on the extraction of V was shown in Fig. 5. It demonstrated that the extraction ratio decreased with the increase in R ranging from 4 to 12. The R on a scale of 4–10, the extraction ratio of V is higher than 99.23%. At R of 12, the extraction

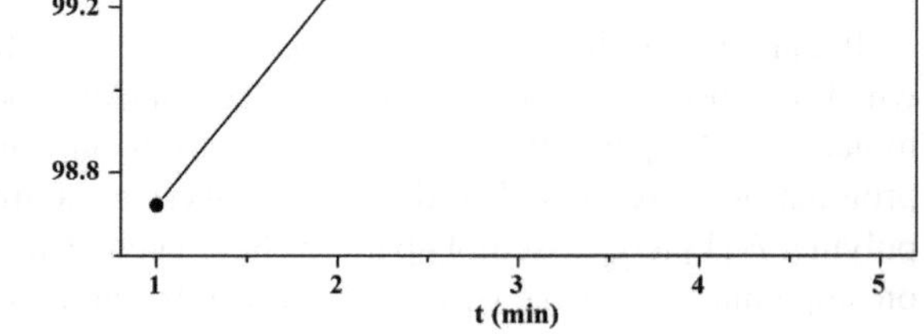

Fig. 4 Effect of contacting time on the extraction ratio of V

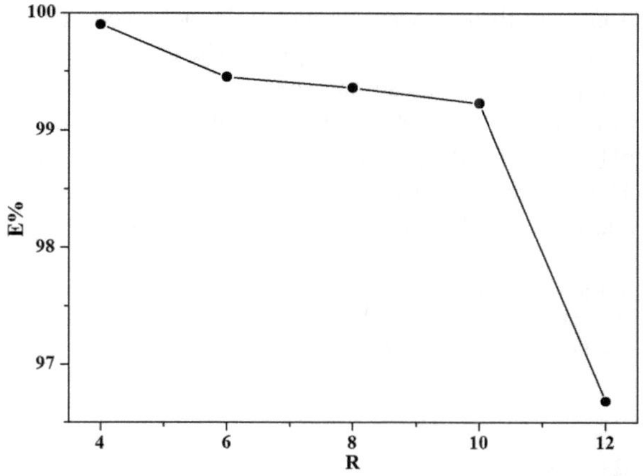

Fig. 5 Effect of R on the extraction ratio of V

Table 1 Chemical compositions of the APV-precipitated wastewater

	V (mg/L)	P (mg/L)	Si (mg/L)	Ca (mg/L)	Al (mg/L)
APV-precipitated wastewater	83.61	0.92	4.25	63.43	9.17

ratio of V dropped to 96.68%. As you can see from Fig. 4, the microemulsion has huge extraction capacity of V. In order to keep the extraction ratio of V, R should be less than 10.

Application

The pH of the APV-precipitated wastewater was controlled at 3. Chemical compositions of the APV-precipitated wastewater are listed in Table 1. We were applying the Aliquat 336/isoamyl alcohol/n-heptane/NaCl microemulsion system to extract V in APV-precipitated wastewater. Extraction experiments were performed using Aliquat 336 (0.112 mol/L) microemulsion with the phase ratio (A/O) designed as 4/1. The results showed that the proposed method has high extraction ratio, in spite of the V concentration is very low in APV-precipitated wastewater. The extraction ratio of V can be achieved 99.74% during the extraction process.

Conclusions

Aliquat 336/isoamyl alcohol/n-heptane/NaCl microemulsion system was developed to extract V in aqueous solution. Experimental results indicated that the extraction equilibrium was immediately achieved, and the extracted compound was $[R_3CH_3N^+]_2[H_4V_{10}O_{28}^{2-}]$ at pH 3. The extraction mechanism of V with Aliquat 336 was proved to be an anion exchange reaction. For synthetic solution, the V extraction ratio can reach 99.90%. The ratio of V extraction can be attained 99.74%, when the pH value of APV-precipitated wastewater is at 3, Aliquat 336 (0.112 mol/L), R = 4. The experiments proved that the microemulsion system has high ratio of V extraction, and it is one of the effective ways to dispose APV-precipitated wastewater.

Acknowledgements This work was financially supported by National Natural Science Foundation of China (Nos. 51474041, 51674051), and Fundamental Research Funds for the Central Universities of China (106112015CDJZR465505, cqu2017hbrc1B08).

References

1. Yao Z (2005) Review of vanadium processing in China. Eng Sci 3:58–62
2. Moskalyk RR et al (2003) Processing of vanadium: a review. Miner Eng 16(9):793–805
3. Yang Z et al (2014) Leaching kinetics of calcification roasted vanadium slag with high CaO content by sulfuric acid. Int J Miner Process 133:105–111
4. Xu Y et al (2011) Study on recovery of vanadium from low-vanadium Slag by Calcified roasting methods. Adv Mater Res 284–286:2127–2130
5. Fang HX et al (2012) Effective chromium extraction from chromium-containing vanadium Slag by Sodium roasting and water leaching. ISIJ Int 52(11):1958–1965
6. Li M et al (2016) Effective extraction of vanadium and chromium from high chromium content vanadium Slag by sodium roasting and water leaching. Materials Science Forum, Korea
7. Lima RS et al (2014) Microemulsification: an approach for analytical determinations. Anal Chem 86(18):9082–9090
8. Shen L et al (2014) Microemulsion extraction of monascus pigments from nonionic surfactant using high polarity of Diethyl Ether as excess oil phase. Sep Sci Technol 49(15):2346–2351
9. Yu T et al (2012) Microemulsion extraction of gold(III) from hydrochloric acid medium using ionic liquid as surfactant and extractant. Ind Eng Chem Res 51(50):16438–16443
10. Lu W et al (2011) Extraction of gold(III) from hydrochloric acid solutions by CTAB/n-heptane/iso-amyl alcohol/Na_2SO_3 microemulsion. J Hazard Mater 186(2–3):2166–2170
11. Lou Z et al (2015) Selective extraction and separation of Re(VII) from Mo(VI) by TritonX-100/N235/iso-amyl alcohol/n-heptane/NaCl microemulsion system. Hydrometallurgy 157:199–206
12. Zheng Y et al (2016) Extraction of palladium (II) by a silicone ionic liquid-based microemulsion system from chloride medium. Sep Purif Technol 169:289–295
13. Jie F et al (2018) Study on extraction of cobalt(II) by sodium laurate/pentan-1-ol/heptane/NaCl microemulsion system. J Radioanal Nucl Chem 315(3):581–593
14. Lu J et al (2011) Microemulsion extraction separation and determination of aluminium species by spectrofluorimetry. J Hazard Mater 185(2):1107–1114
15. Zeng L et al (2009) Extraction of vanadium from the leach solution of stone coal using ion exchange resin. Hydrometallurgy 97(3):194–197
16. Nakamura S et al (2001) Hydrogen-bonded aggregates of protonated decavanadate anions in their tetraalkylammonium salts. J Chem Soc Dalton Trans 28(4):472–480

A Novel Approach for Pre-concentrating Vanadium from Stone Coal

Daya Wang and Baijun Yan

Abstract Stone coal is an important vanadium-bearing resource in China and accounts for about 87% of the domestic reserves of vanadium. But the currently developed techniques to extract vanadium from stone coal are all confronted with environmental problems. In order to solve those problems, an innovative process for pre-concentrating vanadium from stone coal was designed. Firstly, the environment-friendly Fe_2O_3 is selected as a capturer of vanadium. Secondly, the reducing nature of stone coal is utilized to promote the formation of magnetic V-rich phase, Fe_2VO_4. Thirdly, a high-grade vanadium concentrate is obtained by magnetic separation method. The feasibility of this design was confirmed using industrial stone coal at a laboratory scale. When 10 mass% Fe_2O_3 relative to stone coal was added and roasted at 1200 °C for 3 h, a high-grade concentrate with vanadium content of 14% and recovery rate of 92% was recovered.

Keywords Vanadium · Stone coal · Reducing roast · Pre-concentration
Magnetic separation

Introduction

Vanadium is an important transition metal element and plays a vital role in several industries, including steel-making, aerospace engineering, catalyzer and vanadium-REDOX storage battery [1–3]. Vanadium–titanium magnetite and Vanadium-bearing stone coal are major vanadium resources in China; nearly 87% of vanadium is reserved in stone coal ore [4]. Therefore, to extract vanadium from stone coal effi-

D. Wang · B. Yan (✉)
School of Metallurgical and Ecological Engineering, University of Science and Technology Beijing, 100083 Beijing, P. R. China
e-mail: baijunyan@ustb.edu.cn

D. Wang
e-mail: dayawangustb@126.com

G. Azimi et al. (eds.), *Rare Metal Technology 2019*, The Minerals, Metals & Materials Series, https://doi.org/10.1007/978-3-030-05740-4_5

ciently and environment-friendly is of significance for the sustainable supply of vanadium in China as well as in the world.

Stone coal is a sedimentary shale, which is formed by long-term saprofication and coalification under reducing conditions by organisms such as bacteria, algae, sediments in shallow seas, lagoons and bays [4]. It mainly consists of mica group minerals, as well as a small amount of calcite, pyrite and carbonaceous [5]. The vanadium grade in stone coal is generally in the range of 0.13–1.2 mass%. The V^{3+} and a fraction of V^{4+} substitute for Al^{3+} from the dioctahedral structure in mica minerals as an isomorphism, while some V^{4+} and a small amount of V^{5+} occur in an absorbed state on the surface of kaoline and pyrite [6, 7]. The dispersive occurrence of vanadium makes it difficult to pre-concentrate and extract vanadium from stone coal. Firstly, vanadium cannot be pre-concentrated effectively by the conventional flotation, gravity and magnetic separation methods. Secondly, vanadium cannot be leached directly since the trivalent vanadium, V^{3+}, embedded in the crystal lattice of mica minerals is insoluble in acid and water [8–10].

In order to extract vanadium from stone coal, many improved techniques have been developed since the 1970s, such as the blank roasting-acid leaching, the calcified roasting-carbonate leaching and the low salt roasting-cyclic oxidation technique [5, 11–13]. In those vanadium extraction technologies, the V^{3+} is converted to V^{5+} or V^{4+} compounds, and then, the soluble vanadate is leached. But such principle is against the mineral properties of stone coal. Firstly, stone coal is a kind of reductive ore, the oxidizing roast will cause low roasting efficiency, high-energy consumption and high emission. Secondly, the ore processing capacity is quite high in the leaching process without pre-concentration of vanadium. The high throughput of stone coal has caused many serious problems, such as huge amount of feedstock, excessive acid and large amount of the resulting effluents. Hence, in order to extract vanadium from stone coal efficiently and greenly, an innovative process to pre-concentrate vanadium should be designed urgently [14, 15].

In the present process, the environment-friendly Fe_2O_3 was used as an additive, which can capture vanadium in the stone coal. During the reducing roast, the magnetic V-rich Fe_2VO_4 phase was formed. Subsequently, the V-rich phase was separated by magnetic separation method.

Details of the Process

The sequence of the process can be summarized as three key steps. First, the stone coal ore is ground with Fe_2O_3 and carbon, and then pelleted. Second, the pellets are roasted under reducing conditions. Third, the roasted pellets are ground into powder, and the V-rich phase is separated by magnetic method. It should be noted that the quantity of carbon additive is determined by the contents of Fe_2O_3 and the carbon in stone coal.

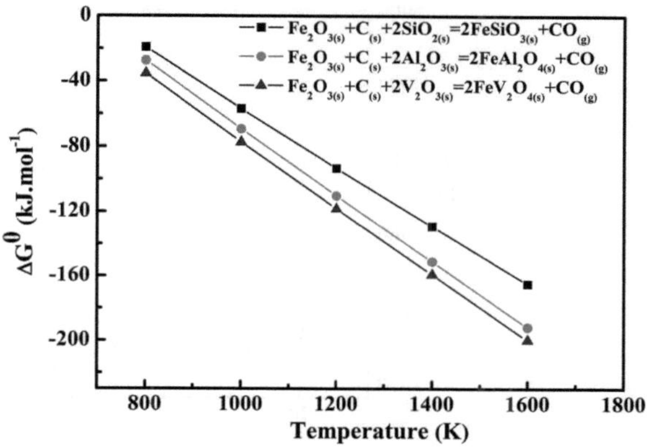

Fig. 1 Changes of standard Gibbs energy of the reactions between iron oxide and V_2O_3, SiO_2, as well as Al_2O_3

The dominant minerals in stone coal are mica, which are composed mainly of SiO_2 and Al_2O_3, the chemical affinities of FeO to V_2O_3, SiO_2 and Al_2O_3 were compared, respectively. The changes of standard Gibbs energy of the reactions during reduction roasting are presented in Fig. 1. It can be seen that the change of Gibbs energy to form FeV_2O_4 is the most negative one, which means that it is more prone to be formed.

Experimental

Characterization of Stone Coal Ore

The raw stone coal ore used in the present study was taken from Yichang City in Hubei province of China. It was crushed, grounded and sieved by a standard test sieve whose pore size was 0.074 mm in diameter. Then, X-ray diffraction analyzer (Bruker D8 Advance) and X-ray fluorescence analyzer (Shimadzu XRF-1800) were used for mineralogical and chemical analysis, respectively.

The XRD pattern is shown in Fig. 2. As seen, it mainly consists of quartz, dolomite, roscoelite, illite, orthoclase and pyrite. Combined with the SEM measurements, it can be concluded that vanadium mainly occurs in illite and roscoelite, and a small amount of vanadium can be found in pyrite and orthoclase mineral. The contents of the elements measured by XRF are presented in Table 1. Furthermore, the total content of carbon determined by the combustion method is about 5.7 mass%, and 5 mass% in the 5.7 mass% comes from carbonate, and the rest is amorphous carbon.

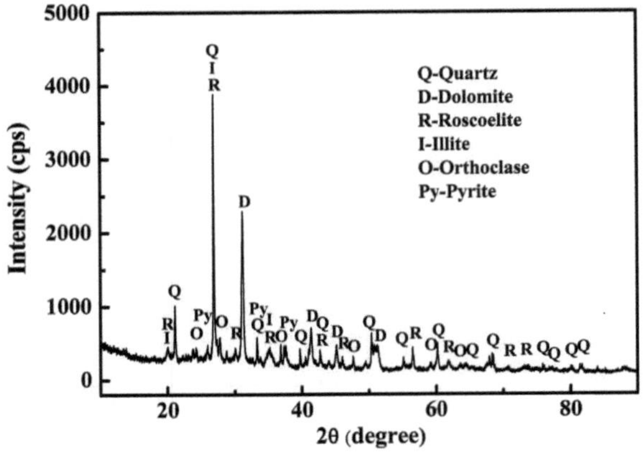

Fig. 2 XRD pattern of raw stone coal

Table 1 XRF analysis results of main compositions in raw stone coal (mass, %)

Composition	Concentration (Mass, %)	Composition	Concentration (Mass, %)
V	0.72	Ca	9.87
Fe	2.71	Mg	4.46
S	1.58	K	4.59
Si	20.64	Ba	0.65
Al	6.26	Ti	0.33

Roasting of Pellets

The stone coal powder was mixed with Fe_2O_3 and carbon powders by using a planetary ball milling. The molar ratio of Fe_2O_3 to carbon powder was unity. The content of Fe_2O_3 relative to the stone coal changed from 5.0, 8.0, up to 10% in different experimental runs. The mixed powder was pelletized, the diameter of the pellets was controlled to be 6–8 mm, and then, these pellets were dried at 80 °C in a drying oven. The dried pellets were roasted at 1000, 1100 and 1200 °C for 3 h in Ar atmosphere. The roasting was conducted in a vertical furnace shown in Fig. 3.

The operation procedure of a typical run was as follows: The crucible containing the pellet sample was placed in the water-cooled quenching chamber before heated. After the whole system was carefully sealed, the reaction tube was evacuated using a vacuum pump and then filled with argon at least 3 times. The furnace was then heated to the desired temperatures with a heating rate of 3 K/min. When the desired temperature was achieved, the crucible containing the sample was introduced into the even temperature zone. After being held for 3 h, the crucible was lifted to the quenching chamber, and then the roasted pellets were taken out. The mineral compositions of roasted pellets were characterized by XRD, the morphology and composition analy-

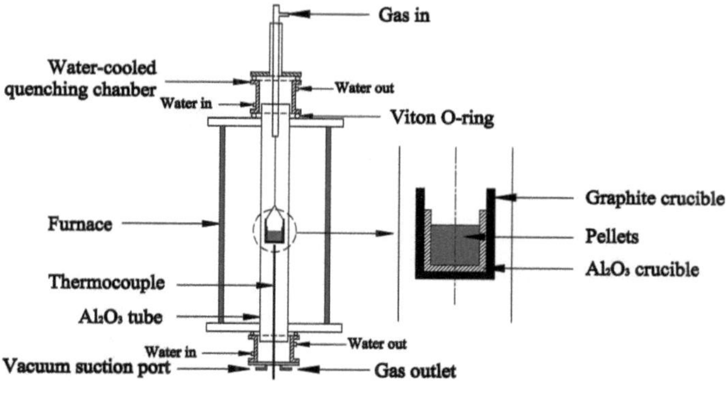

Fig. 3 Schematic drawing of the furnace for roasting

sis were conducted by SEM-EDS, and the magnetic property was determined using the magnetometer (VSM, 9600-1 LDJ, USA).

Magnetic Separation of the Vanadium-Rich Phase

The V-rich phase formed during reduction roasting can be separated by magnetic separation method. Firstly, the roasted pellets were ground into powder of which the particle size was all less than 0.065 mm. The powder was then dispersed in deionized water to form uniform pulp. Secondly, a 0.5 T magnet covered by a copper shell was dipped and moved randomly in the pulp. The magnetic minerals were attracted on the outer surface of the copper shell.

The magnet and the copper shell were taken out of the pulp, and the magnetic substance attracted on the copper shell were flushed by deionized water. The minerals detached with deionized water were denoted as weak magnetic minerals. The minerals still attracted by the magnet are referred as strong magnetic concentrates, which were collected by removing the magnet out of the copper shell. The minerals remaining in the pulp were filtered, and the nonmagnetic tailings were obtained. The collected strong magnetic, weak magnetic and nonmagnetic minerals were dried and then characterized and analyzed by XRD, SEM-EDS, respectively. In addition, ICP-AES was adopted to measure the content of V.

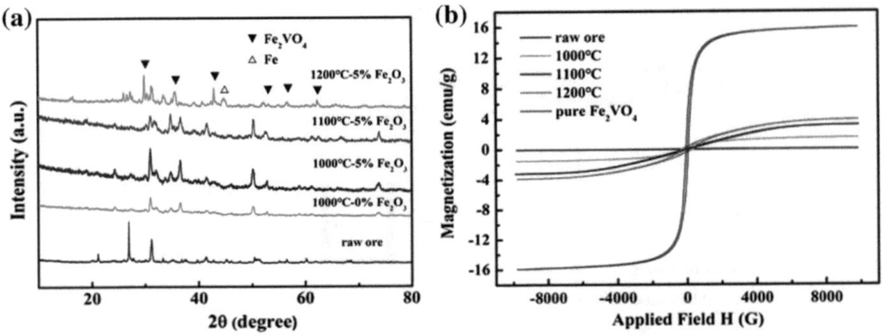

Fig. 4 XRD patterns (**a**) and magnetic hysteresis loops (**b**) of the pellets roasted with 5 mass% Fe_2O_3 additive at 1000, 1100 and 1200 °C

Results and Discussion

Effect of Roasting Temperature on the Formation of V-Rich Phase

One kind of pellet prepared by adding 5 mass% Fe_2O_3 was used to investigate the effect of roasting temperature on the formation of V-rich phase. The pellets were roasted for 3 h in Ar atmosphere at 1000, 1100 and 1200 °C, respectively. An alumina crucible was used to hold the pellets. Besides, a graphite holder was not only adopted to support the Al_2O_3 crucible, but also used to remove the trace amounts of oxygen in argon (Ar) atmosphere.

The phases in the roasted pellets were identified firstly by XRD; the patterns are presented in Fig. 4a. For the sake of comparison, the XRD patterns of the pellets without Fe_2O_3 additive and the raw stone coal ore are listed together. It can be seen that when the roasting temperature was below 1100 °C, it is only the decomposition of mica minerals that can be distinguished. Also, the XRD patterns of the pellets in which Fe_2O_3 was added or not are very similar to each other at 1000 °C. When the roasting temperature is reached to 1100 °C, some small peaks belonging to Fe_2VO_4 are identified. At 1200 °C, the strongest peaks in the roasted samples are attributed to be Fe_2VO_4, which means that Fe_2VO_4 is the dominant phase.

In order to investigate the micro-distributions of Fe and V in the pellets roasted at 1200 °C, SEM-EDS measurements were performed. A representative micrograph of the pellets with 5 mass% Fe_2O_3 was shown in Fig. 5, and the corresponding EDS analyses for element Fe and V are presented together. It can be seen clearly that Fe and V exhibit similar distributing behavior, and a phase composed of Fe and V with diameter of 30 μm was observed. These results indicate that the iron oxide is an effective medium to pre-concentrate vanadium. A stable V-rich phase, Fe_2VO_4 can be formed under the reducing atmosphere at 1200 °C.

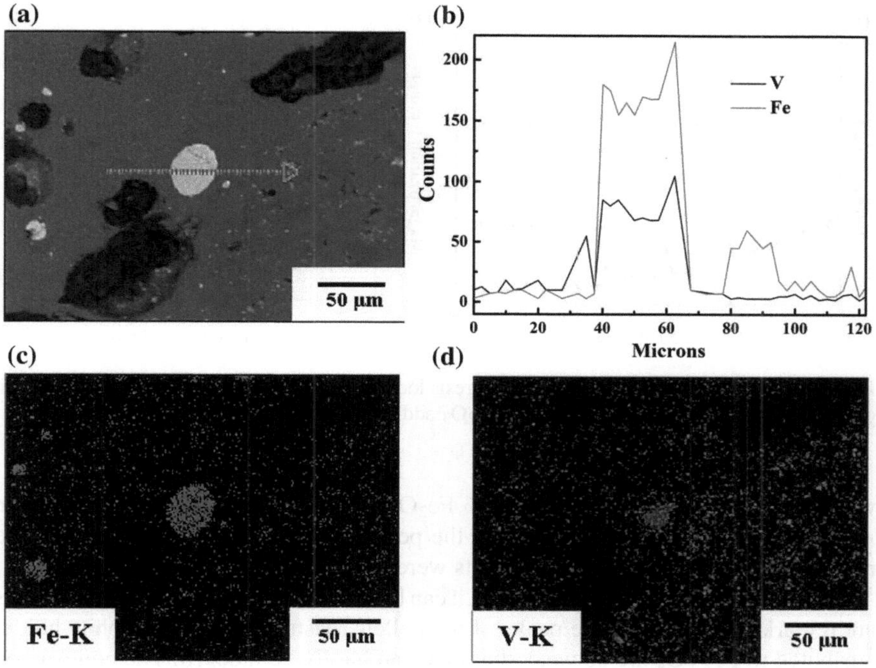

Fig. 5 SEM image of the pellets with 5 mass% Fe_2O_3 at 1200 °C (**a**), the line scan along the arrow in the SEM image (**b**) and area scans (**c**) and (**d**) for element Fe and V

The magnetic hysteresis loops of the pellets are presented together with that of pure Fe_2VO_4 compound in Fig. 4b. It can be seen firstly that the pure Fe_2VO_4 compound exhibits strong magnetic property, which means that Fe_2VO_4 is an ideal V-rich phase from the point of view of magnetic separation. Also, the magnetism of the roasted pellets increases with the increasing temperature. Such trend is consistent with the amount of Fe_2VO_4 phase measured by XRD. Also, according to the SEM measurements, no magnetite was found within the roasted samples. Considering the small amount of Fe, it can be indicated that the magnetization is mainly contributed by the Fe_2VO_4 phase, which means that the vanadium in stone coal can be concentrated as a magnetic Fe_2VO_4 phase during the reducing roast.

Effect of Oxygen on the Formation of V-Rich Phase

In this part, the effect of oxygen in the atmosphere on the concentrating of vanadium in stone coal by Fe_2O_3 was preliminarily investigated. There are two different experimental groups. In the first group, the alumina crucible was held by a graphite crucible during roasted. In the second one, the graphite holder out of the alumina crucible

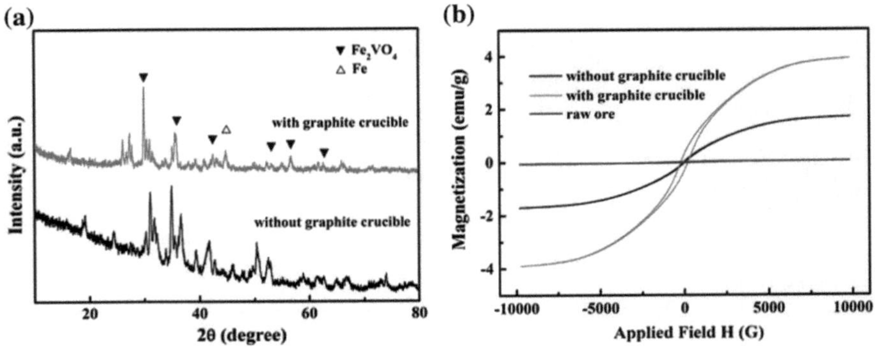

Fig. 6 XRD patterns (**a**) and magnetic hysteresis loops (**b**) of the pellets roasted with and without graphite holder at 1200 °C with 5 mass% Fe_2O_3 additive

was absent. The pellets with 5 mass% Fe_2O_3 additive were roasted at 1200 °C in Ar atmosphere for 3 h. After roasted, the pellets were also analyzed by XRD and magnetic measurements, and the results were shown in Fig. 6.

From XRD pattern shown as Fig. 6a, it can be seen clearly that the diffraction peaks are remarkably different, the peaks of the pellets roasted without a graphite holder attributed to Fe_2VO_4 are so weak that they can hardly be indexed. The comparison of the magnetic property is presented in Fig. 6b. The magnetization of the pellets roasted with graphite holder is stronger than that roasted without a graphite holder, which means that the amount of magnetic substances in the pellets roasted with a graphite holder are higher.

From the above results of XRD and magnetic measurements, it can be concluded preliminarily that oxygen is a sensitive factor to the formation of Fe_2VO_4 phase, and the lower oxygen partial pressure may be more advantageous to the concentrating of vanadium in stone coal by Fe_2O_3.

Effect of the Quantity of Fe_2O_3 on the Formation of V

To determine the appropriate quantity of Fe_2O_3, the enrichment and the recovery rate of vanadium were investigated under different Fe_2O_3 amount ranging from 5 mass% to 10 mass%. The XRD patterns and the magnetic measurements of the roasted pellets are shown in Fig. 7. It can be revealed that the more Fe_2O_3 added, the stronger the magnetization of the roasted pellets became. This indicates that the amount of the magnetic substances in the roasted sample is influenced by the adding amount of Fe_2O_3, thereby the enrichment and the recovery rate of vanadium will possibly be affected.

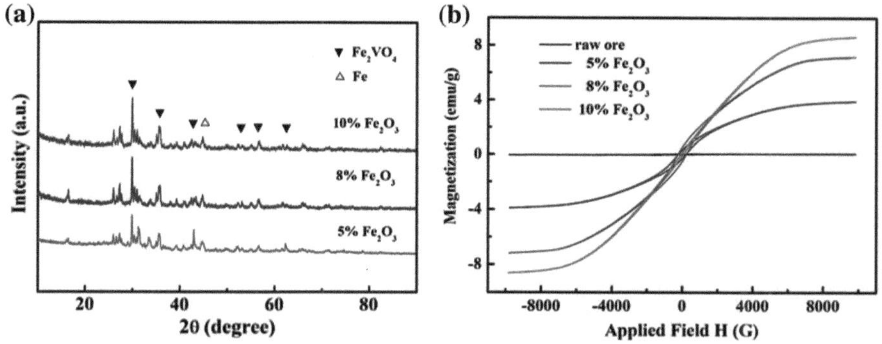

Fig. 7 XRD patterns (**a**) and magnetic hysteresis loops (**b**) of the pellets roasted at 1200 °C with various quantities of Fe_2O_3 added

To determine the recovering rate of vanadium quantitatively, the magnetic phases in the roasted pellets were separated. Then, the contents of vanadium in the strong magnetic concentrates, the weak magnetic minerals, and the nonmagnetic tailings were analyzed by ICP-AES, and the results are shown in Table 2. From Table 2, it can be found firstly that the vanadium content in the strong magnetic concentrates is much higher than that in raw stone coal ore, the weak magnetic minerals and the tailings. In addition, with the increasing additive content of Fe_2O_3, the content and recovery rate of vanadium in concentrate are increased. For the pellets with 10 mass% Fe_2O_3 added, the content of vanadium in the concentrates is more than 14 times higher than that in stone coal ore, and the recovery rate of vanadium is also higher than 90%. These results indicate that vanadium in stone coal can be efficiently enriched with 10 mass% Fe_2O_3 additive.

The morphology, mineral compositions and the phases of the concentrates were further studied by SEM-EDS and XRD. From the XRD pattern shown in Fig. 8a, it can be seen clearly that the concentrates mainly consist of Fe_2VO_4 and a small amount of Fe. The micrograph of the concentrates is shown in Fig. 8b. Three different kinds of minerals, Fe_2VO_4, Fe and small amount of aluminosilicate, are also detected.

Conclusions

In order to extract vanadium from stone coal effectively and greenly, an innovative process for pre-concentrating vanadium was designed and confirmed on a laboratory scale. The low-cost and environment-friendly Fe_2O_3 is used to be an appropriate pre-concentrating medium. Based on the reducing nature of stone coal, a V-rich inverse spinel phase Fe_2VO_4 is formed during the reduction roasting, then separated by magnetic method. After the optimized conditions for the formation of V-rich magnetic phase, Fe_2VO_4 are determined experimentally. A kind of high-grade vanadium con-

Table 2 Contents of vanadium in concentrate and tailings analyzed by ICP-AES

Roasted pellets			Strong magnetic concentrate		Weak magnetic concentrate		Tailings		Recovering rate of V (%)
Mass (g)	Fe_2O_3 (mass%)	V (mass%)	Mass (g)	V (mass%)	Mass (g)	V (mass%)	Mass (g)	V (mass%)	
39.90	5	0.81	2.15	9.00	4.75	1.22	32.46	0.01	59.9
39.82	8	0.92	2.36	10.30	3.63	1.79	33.10	0.09	66.4
40.18	10	1.04	2.65	14.49	3.01	0.35	32.89	0.06	91.9

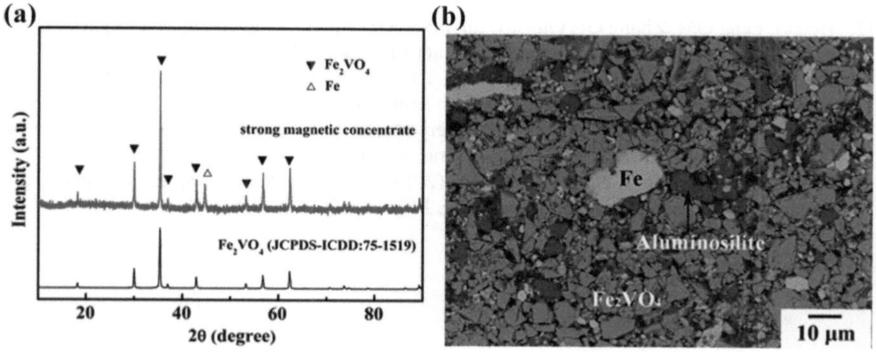

Fig. 8 XRD pattern (**a**) and SEM image (**b**) of the concentrate

centrate containing V higher than 14 mass% was obtained, and more than 92% of the total vanadium in stone coal ore was recovered.

Acknowledgements The financial support on the projects 51534001 and 51474009 from the National Natural Science Foundation of China is gratefully acknowledged.

References

1. Moskalyk RR, Alfantazi AM (2003) Processing of vanadium: a review. Miner Eng 16(9):793–805
2. Yamamura T, Wu XW, Ohta S, Shirasaki K, Sakuraba H, Satoh I, Shikama T (2011) Vanadium solid-salt battery: solid state with two redox couples. J Power Sources 196(8):4003–4011
3. Aboelfetoh EF, Peistchnig R (2014) Preparation, characterization and catalytic activity of MgO/SiO_2 supported vanadium oxide based catalysts. Catal Lett 144(1):97–103
4. Dai SF, Ren DY, Chou CL, Finkelman RB, Seredin VV, Zhou YP (2012) Geochemistry of trace elements in Chinese coals: a review of abundances, genetic types, impacts on human health, and industrial utilization. Int J Coal Geol 94:3–21
5. Zhang YM, Bao SX, Liu T, Chen TJ, Huang J (2011) The technology of extracting vanadium from stone coal in China: history, current status and future prospects. Hydrometallurgy 109(1–2):116–124
6. Wang MY, Xiang XY, Zhang LP, Xiao LS (2008) Effect of vanadium occurrence state on the choice of extracting vanadium technology from stone coal. Rare Met 27(2):112–115
7. Hu YJ, Zhang YM, Bao SX, Liu T (2012) Effects of the mineral phase and valence of vanadium on vanadium extraction from stone coal. Int J Miner Metall Mater 19(10):893–898
8. Liu X, Zhang YM, Liu T, Cai ZL, Sun K (2016) Pre-concentration of vanadium from stone coal by gravity using fine mineral spiral. Minerals 6(3):82–92
9. Zhao YL, Zhang YM, Liu T, Chen TJ, Bian Y, Bao SX (2013) Pre-concentration of vanadium from stone coal by gravity separation. Int J Miner Process 121(121):1–5
10. Wang L, Sun W, Liu RQ, Gu XC (2014) Flotation recovery of vanadium from low-grade stone coal. Trans Nonferrous Met Soc China 24(4):1145–1151
11. Li MT, Wei C, Fan G, Li CX, Deng ZG, Li XB (2009) Extraction of vanadium from black shale using pressure acid leaching. Hydrometallurgy 98(3–4):308–313

12. He DS, Feng QM, Zhang GF, Ou LM, Lu YP (2007) An environmentally friendly technology of vanadium extraction from stone coal. Miner Eng 20(12):1184–1186
13. Zhu XB, Zhang YM, Huang J, Liu T, Wang Y (2012) A kinetics study of multi-stage counter-current circulation acid leaching of vanadium from stone coal. Int J Miner Process 114–117:1–6
14. Yang HF, Jing LL, Zhang BG (2011) Recovery of iron from vanadium tailings with coal-based direct reduction followed by magnetic separation. J Hazard Mater 185(2–3):1405–1411
15. Li J, Zhang YM, Liu T, Huang J, Bao SX (2014) A methodology for accessing cleaner production in the vanadium extraction industry. J Cleaner Prod 84:598–605

Study on the Roasting Mechanism of Vanadium–Chromium Slag with Sodium Hydroxide

Minmin Lin, Chengjie Wang, Bing Xie and Hong-Yi Li

Abstract It has been most widely used to extract vanadium and chromium by sodium roasting–water extraction. However, the current researches about this evolution law only stay at the macro-level. Therefore, it is especially important to study the mechanism of oxidation and sodium consumption in the process of sodium roasting. In this study, sodium hydroxide was used as a sodium additive for roasting experiments. The oxidation and sodiumation processes were systematically studied, providing a theoretical basis for industrial extraction of vanadium and chromium. The results show that the oxidation and sodiumation temperature range of V^{3+} and Cr^{3+} is from 425.8 to 882.8 °C. The roasting temperature of 800 °C is the optimum conditions. Under this condition, vanadium spinel is converted into $NaVO_3$, while chromium spinel is converted into Na_2CrO_4 and $CaCrO_4$.

Keywords Vanadium–chromium slag · Sodium hydroxide · Roasting mechanism

Introduction

Vanadium is an important strategic resource widely used in the steel, aerospace and chemical industries due to its excellent physical properties [1]. The vanadium slag, which is obtained from smelting of vanadium and titanomagnetite, is the most important raw material for extraction of vanadium. At present, the main industrial process for vanadium extraction from vanadium slag is sodium roasting–water leaching extraction technique [2–4]. It is well known that oxidation roasting is a key stage in the whole process of vanadium extraction. The objective of salt roasting is to convert vanadium to water-soluble sodium salt [5–7].

The current researches about this evolution law only stay at the macro-level. Many studies have focused on the vanadium recovery from vanadium slag through optimizing the process conditions to develop an economical and effective production

M. Lin · C. Wang · B. Xie · H.-Y. Li (✉)
College of Materials Science and Engineering, Chongqing University, Chongqing 400044, China
e-mail: hongyi.li@cqu.edu.cn

G. Azimi et al. (eds.), *Rare Metal Technology 2019*, The Minerals, Metals & Materials Series, https://doi.org/10.1007/978-3-030-05740-4_6

route [8, 9]. Studies on oxidation process indicate that the conversion rate of vanadium is strongly dependent on the roasting temperature, roasting time and additive content [10–12]. However, as far as we know, there are few reports on the mechanism of oxidation and sodium consumption in the process of sodium roasting.

Therefore, in this work, the oxidation mechanism of vanadium and chromium and the consumption mechanism of sodium are proposed by the oxidation roasting experiment, which was conducted at different temperatures in the presence of NaOH. The changes in vanadium–chromium slag were studied by XRD and TG-DSC techniques.

Experimental

Materials

The vanadium–chromium slag used in the experiment was obtained by a duplex process from Panzhihua Steel & Iron (Group) Corporation, Sichuan, China. After crushing, magnetic separation and grinding, the materials were screened through 74 μm sieve. The chemical composition of vanadium slag is given in Table 1. The XRD pattern of the vanadium slag is shown in Fig. 1, which mainly consists of spinel phases ((Mn,Mg,Fe) $(Mn,Cr,V,Fe)_2O_4$, Mg_2TiO_4), pyroxene phase ($CaAl_2SiO_6$) and olivine phase (Fe_2SiO_4).

In order to study the distribution of different elements in vanadium chromium slag, the elemental surface scanning analysis of vanadium–chromium slag was carried out by EDS spectrum analyzer. The results are shown in Fig. 2. The first picture is the BSE diagram of vanadium–chromium slag. The diagram shows that the main constituent phases of vanadium–chromium slag are spinel phase, olivine phase and pyroxene phase. According to the results of elemental surface scanning, Mg and Si are mainly concentrated in the olivine phase; Ca and Al are concentrated in the pyroxene phase, which become two unique elements that make up the pyroxene phase. V, Cr and Ti are mainly distributed in the spinel phase. Among them, the V and Ti elements are distributed in all spinel phases, while the Cr element is only present in some spinel phases. The Ti content in the surface region is the highest, that is, titanium distributing in the outermost edge area of the spinel grain, which indicates that the titanium spinels precipitate last, and the reason why the titanium spinel diffraction peak is not detected in the XRD peak is further verified.

Table 1 Chemical composition of vanadium slag (mass fraction, %)

FeO	SiO_2	V_2O_3	TiO_2	Cr_2O_3	MnO	Al_2O_3	MgO	CaO	P_2O_5
38.23	17.65	10.07	9.90	8.27	6.21	4.17	3.50	1.92	0.075

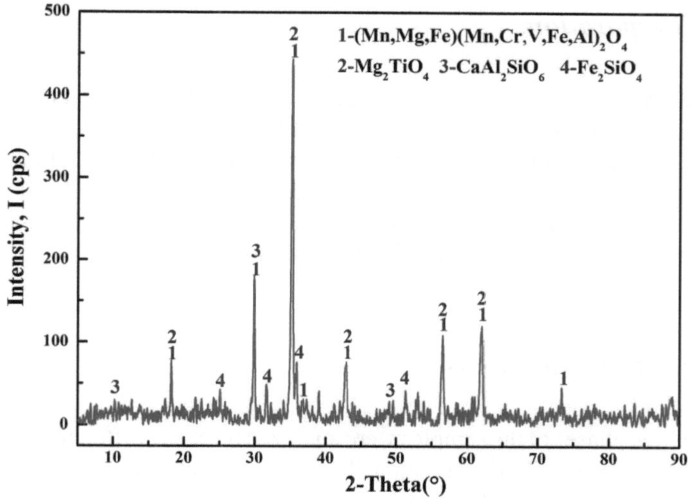

Fig. 1 XRD pattern of vanadium slag

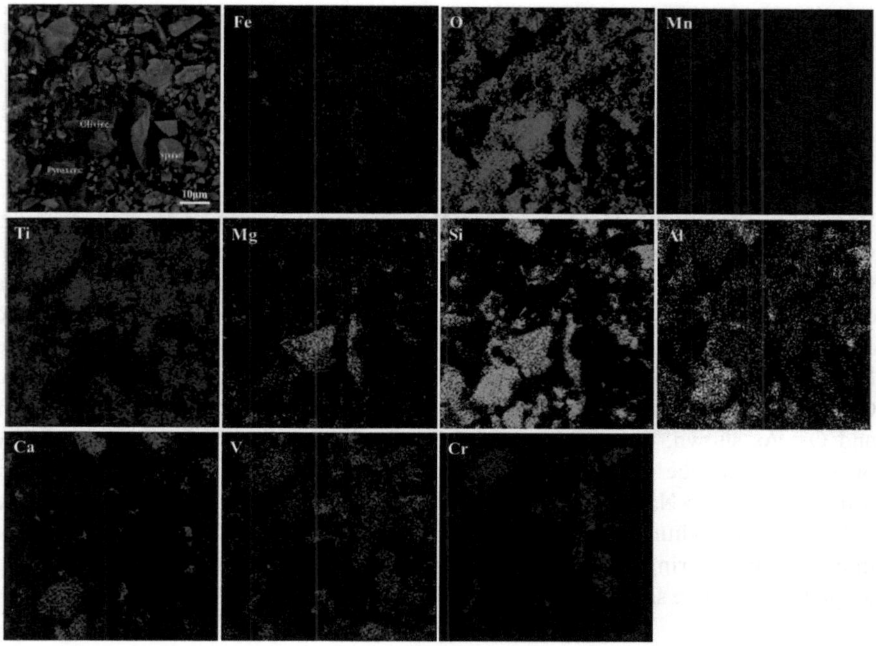

Fig. 2 BSE diagram and element distribution of vanadium–chromium slag

Sample Preparation

The vanadium–chromium slag was thoroughly mixed with analytical reagent NaOH at a mass ratio of 10:3. The mixed samples were put and heated in muffle furnace at required temperature for 3 h. The furnace door was kept half-opened to maintain an oxidizing atmosphere during the roasting process and stirred every 15 min to prevent the samples from sintering. Then, the samples were removed from the furnace, cooled to room temperature, and ground to 74 μm with an agate mortar.

Characterization Methods

The crystalline phases of the samples were analyzed by X-ray diffractometer (Rigaku D/max 2500PC) using Cu Kα radiation. The mixed samples were measured by thermal gravimetry (TG) and differential scanning calorimetry (DSC) at a heating rate of 10 °C/min from room temperature to 1200 °C in air atmosphere.

Results and Discussion

Thermodynamic Analysis of Vanadium Extraction

Vanadium–chromium slag has a complex chemical composition and contains various mineral phases. Currently, only the reaction between the main phase in vanadium–chromium slag and NaOH is considered. The main chemical reactions during roasting are summarized in Table 2. The data required for thermodynamic calculations come from the existing thermodynamic data and the database in the FactSage 6.0 thermodynamic software package (Thermfact Ltd.-CRCT, Montréal, Canada). Figure 3 shows the reaction trend of the main phase reacting with NaOH and O_2. As shown, the reactions can be carried out spontaneously. The stability of the high-valence vanadate salt formed by the reaction of FeV_2O_4 with NaOH follows $NaVO_3 > Na_4V_2O_7 > Na_3VO_4$. It also can be seen from Fig. 3 that the oxidation and sodiumization ability of FeV_2O_4 is stronger than that of $FeCr_2O_4$, indicating that during the roasting process, the vanadium is more easily converted to the high-valence soluble sodium vanadate salt.

Table 2 Main chemical reactions during sodium roasting

Reaction	No.
$\frac{1}{2}$ $FeAl_2O_4$ + NaOH + $\frac{1}{8}$ O_2 = $NaAlO_2$ + $\frac{1}{4}$ Fe_2O_3 + $\frac{1}{2}$ H_2O	(1)
$\frac{1}{2}$ $FeCr_2O_4$ + NaOH + $\frac{1}{8}$ O_2 = $\frac{1}{2}$ Na_2CrO_4 + $\frac{1}{4}$ Fe_2O_3 + $\frac{1}{2}$ H_2O	(2)
$\frac{1}{4}$ $FeCr_2O_4$ + NaOH + $\frac{7}{16}$ O_2 = $\frac{1}{2}$ Na_2CrO_4 + $\frac{1}{8}$ Fe_2O_3 + $\frac{1}{2}$ H_2O	(3)
$\frac{1}{6}$ FeV_2O_4 + NaOH + $\frac{5}{24}$ O_2 = $\frac{1}{3}$ Na_3VO_4 + $\frac{1}{12}$ Fe_2O_3 + $\frac{1}{2}$ H_2O	(4)
2FeO + NaOH + $\frac{1}{2}$ O_2 = $NaFeO_2$ + $\frac{1}{2}$ H_2O	(5)
$\frac{1}{4}$ FeV_2O_4 + NaOH + $\frac{5}{16}$ O_2 = $\frac{1}{4}$ $Na_4V_2O_7$ + $\frac{1}{8}$ Fe_2O_3 + $\frac{1}{2}$ H_2O	(6)
$\frac{1}{2}$ Fe_2SiO_4 + NaOH + $\frac{1}{4}$ O_2 = $\frac{1}{2}$ Na_2SiO_3 + $\frac{1}{2}$ Fe_2O_3 + $\frac{1}{2}$ H_2O	(7)
$\frac{1}{2}$ FeV_2O_4 + NaOH + $\frac{5}{8}$ O_2 = $2NaVO_3$ + $\frac{1}{4}$ Fe_2O_3 + $\frac{1}{2}$ H_2O	(8)

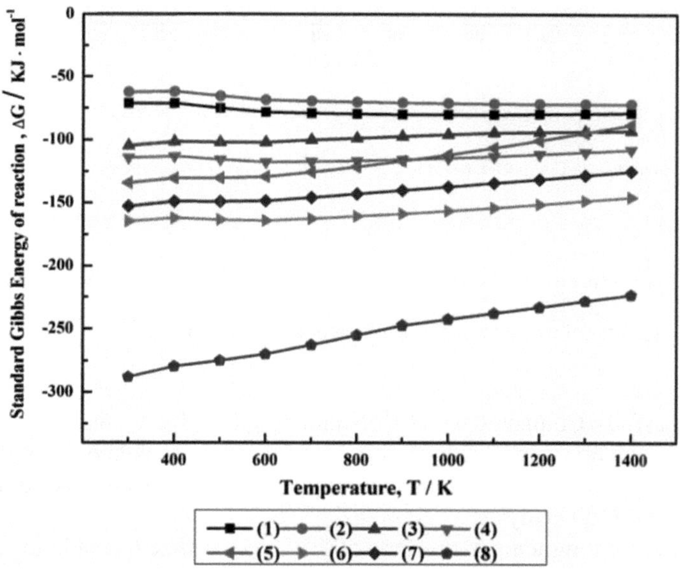

Fig. 3 Variation of standard Gibbs free energy ΔG^θ of several reactions with temperature obtained by thermodynamic calculations using FactSage 6.0 software

XRD Analysis

Figure 4 shows the XRD patterns of the roasted samples at different temperatures. Compared with vanadium slag, the roasted sample at 300 °C presents $Mg_{1.5}VO_4$ and $CrMn_{1.5}O_4$ diffraction peaks. It can be inferred that some V^{3+} and Cr^{3+} in spinel phases have been oxidized to V^{5+} and Cr^{5+}, respectively. At 400 °C, some of the oxidized vanadium reacts with NaOH to form $NaVO_3$, and partial $FeCr_2O_4$ is oxidized

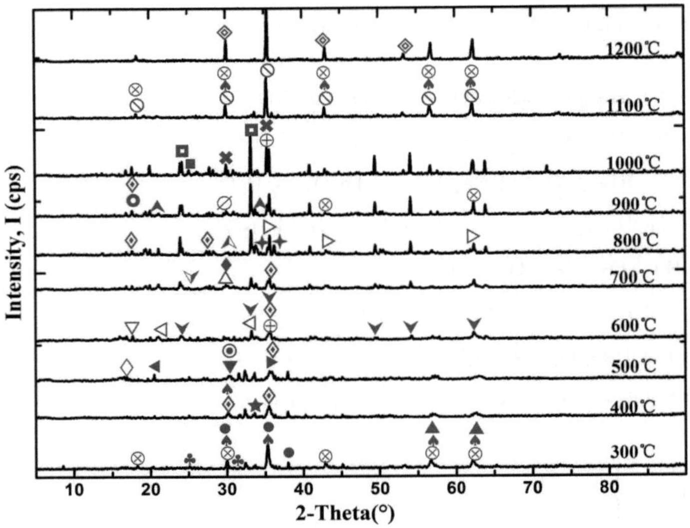

⊗—$Mg_{1.5}VO_4$;♣—Fe_2SiO_4; ▲—$CrMn_{1.5}O_4$; ●—$FeCr_2O_4$; ♠—V_3O_4; ◈—$NaVO_3$; ★—$(Fe_{0.6},Cr_{0.4})_2O_3$;

◊—$CaCr_2O_4$; ▼—Na_2CrO_4; ◉—$Ca_3(CrO_4)_2$; ◀—Na_3VO_4; ▶—$CaVO_3$; ▽—$Na_3V_3O_8$; ◁—$Ca_3Cr_2(SiO_4)_3$;

⊕—$(Fe,Mg)Cr_2O_4$; ⋎—Fe_2O_3; △—NaV_3O_8; ♦—$Ca_3(VO_4)_2$; ⋎—$CaCrO_4$; ✛—Cr_2O_3; ⩓—$NaCr^{+3}Si_2O_6$; ▷—Mg_2VO_4;

∅—$Mn_{1.5}Cr_{1.5}O_4$; ⋏—Na_3CrO_4; ○—$MgVO_3$; ◘—MgV_3O_8; ■—CaV_2O_6; ✖—$Na(V,Cr)Si_2O_6$; ⦸—Mn_3O_4; ◈—Fe_3O_4

Fig. 4 XRD patterns of roasted samples at different temperatures

to $(Fe_{0.6}Cr_{0.4})_2O_3$. Compared with Cr-containing spinel, the V-containing spinel first undergoes oxidation and sodiumation to form sodium vanadate. This result is in good agreement with the previous thermodynamic analysis and also indicates the correctness of XRD analysis.

When the roasting temperature was 500 °C, the oxidized vanadium reacts with NaOH to produce Na_3VO_4. The formation of Na_2CrO_4 was also found in the diffraction peaks, indicating that the oxidation and sodiumation of Cr^{3+} in the chrome spinel occur at this temperature. In addition, $CaAl_2SiO_6$ reacts with spinel phases to form $CaVO_3$ and calcium chromate containing low-valent chromium, such as $Ca_3(CrO_4)_2$ and $CaCr_2O_4$.

As the roasting temperature reaches 600 °C, the characteristic peaks of olivine phases disappear, whereas the diffraction peak of Fe_2O_3 appears, and its relative intensity increases with the increase in roasting temperature. At 700 °C, the sodiumation products of vanadium include NaV_3O_8 and $NaVO_3$. It was also found that $CaAl_2SiO_6$ continued to react with spinel phases to form $Ca_3(VO_4)_2$ and $CaCrO_4$.

At 800 °C, part of $(Fe,Mg)Cr_2O_4$ decomposes to form Cr_2O_3, and at the same time, a part of unoxidized Cr^{3+} reacts directly with NaOH to form $NaCrSi_2O_6$ (minor phase). In addition, the main Cr-containing phase is $CaCrO_4$ and Na_2CrO_4. The sodium vanadate salt at this temperature is the most stable $NaVO_3$, which is also one of the major phases. It indicates that most of the V^{3+} has undergone oxidation and sodiumization reaction to produce a large number of target products.

When the roasting temperature increases to 900 °C, the sample begins to sinter, which slows down the oxidation and sodiumation of V^{3+} and Cr^{3+}. The appearance of $Mn_{1.5}Cr_{1.5}O_4$ and $MgVO_3$ indicates that the oxidation of V^{3+} and Cr^{3+} is not sufficient due to the obstruction of oxygen mass transfer.

As the roasting temperature is further increased, the sintering of the sample becomes more pronounced. When the roasting temperature exceeds 1000 °C, the sodiumation reaction is substantially inhibited, and only, the oxidation reaction of V^{3+} and Cr^{3+} occurs. Main phase includes Mg_2VO_4, $Mg_{1.5}VO_4$, $CrMn_{1.5}O_4$ and V_3O_4. In practical roasting processes, melting of components with low melting points leads to sample sintering when the roasting temperature exceeds 900 °C. Sintering of the sample results in less air gap inside the sample, which hinders the diffusion of oxygen and finally inhibits the oxidation and sodiumation of V^{3+} and Cr^{3+}. Therefore, the excessive roasting temperature is unfavorable for obtaining water-soluble sodium vanadate salt and sodium chromate salt.

TG-DSC Analysis

Figure 5 shows the TG-DSC curves of mixed sample. It can be seen from the TG curve that the mass increment is 1.1% from 25 to 100 °C, which is caused by deliquescent of NaOH. It is also found from the TG curve that the sample continues to gain weight from 425.8 to 882.8 °C. Combining with XRD analysis, the mass increment is attributable to the oxidation of Fe^{2+} in olivine phases and partial Fe^{2+}, Mn^{2+}, Cr^{3+} and V^{3+} in spinel phases. Meanwhile, partial NaOH can react with the above-oxidized products to form a variety of sodium salts. Therefore, this interval (425.8–882.8 °C) can be defined as the oxidation and sodiumation temperature range of V^{3+} and Cr^{3+}. According to the analysis results of XRD, the olivine phase (Fe_2SiO_4) starts to decompose at 300 °C, and the oxidative decomposition of the spinel phase occurs at about 600 °C and reacts with the NaOH. Therefore, exothermic peaks appear at 338.6 and 568.8 °C, respectively.

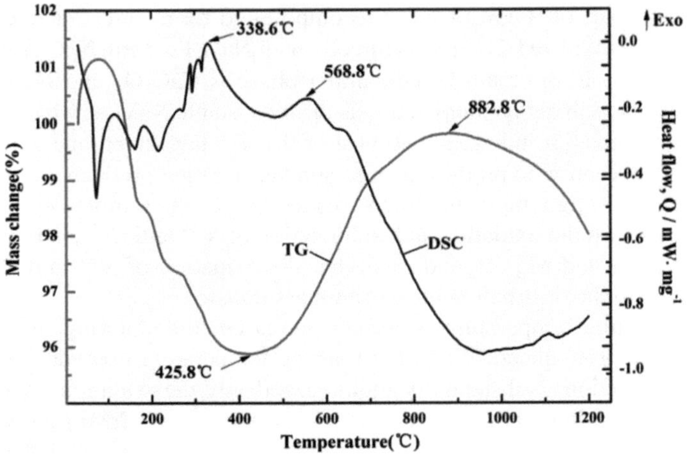

Fig. 5 TG-DSC curves of mixed sample

Conclusions

In this study, the oxidation mechanism of vanadium and chromium and the consumption mechanism of sodium are proposed by the oxidation roasting experiment, which was conducted at different temperatures in the presence of NaOH. The conclusions are summarized as follows: V^{3+} and Cr^{3+} in spinel phases have been oxidized to V^{5+} and Cr^{5+} at 300 °C, respectively. At 400 °C, some of the oxidized vanadium reacts with NaOH to form $NaVO_3$, and some of $FeCr_2O_4$ is oxidized to $(Fe_{0.6}Cr_{0.4})_2O_3$. When the roasting temperature was 500 °C, the oxidized chromium reacts with NaOH to produce Na_2CrO_4. Most of the V^{3+} has undergone oxidation and sodiumization reaction to produce a large number of $NaVO_3$ at 800 °C; the main Cr-containing phase is $CaCrO_4$ and Na_2CrO_4. As the roasting temperature increases to 900 °C, the sample begins to sinter, which slows down the oxidation and sodiumation of V^{3+} and Cr^{3+}. Therefore, the optimum roasting temperature is 800 °C. The oxidation and sodiumation temperature range of V^{3+} and Cr^{3+} is from 425.8 °C to 882.8 °C. When NaOH is used as a sodium salt additive, Na reacts with V, Cr and Si to form various compounds to cause consumption of NaOH during roasting process.

Acknowledgements This work was financially supported by National Natural Science Foundation of China (Nos. 51474041, 51674051) and Fundamental Research Funds for the Central Universities of China (106112015CDJZR465505, cqu2017hbrc1B08).

References

1. Moskalyk RR, Alfantatazi AM (2003) Processing of vanadium: a review. Miner Eng 16(9):793–805
2. Li M, Xiao L, Liu JJ et al (2016) Effective extraction of vanadium and chromium from high chromium content vanadium slag by sodium roasting and water leaching. Mater Sci Forum 144–148
3. Fang HX, Li HY, Xie B (2012) Effective chromium extraction from chromium-containing vanadium slag by sodium roasting and water leaching. ISIJ Int 52(11):1958–1965
4. Wang X, Wang M, Fu Z et al (2017) Present status and prospects of vanadium and chromium separation in vanadium extraction from vanadium-chromium slag. Iron Steel Vanadium Titanium 38(6):1–5
5. Gupta CK, Krishnamurty N (1992) Extractive metallurgy of vanadium. Elsevier Science Publishers, Amsterdam, pp 203–380
6. Wen J, Jiang T, Xu Y et al (2018) Efficient separation and extraction of vanadium and chromium in high chromium vanadium slag by selective two-stage roasting-leaching. Metall Mater Trans B 49(3):1471–1481
7. Ning P, Lin X, Wang X et al (2016) High-efficient extraction of vanadium and its application in the utilization of the chromium-bearing vanadium slag. Chem Eng J 301:132–138
8. Li HY, Fang HX, Wang K et al (2015) Asynchronous extraction of vanadium and chromium from vanadium slag by stepwise sodium roasting-water leaching. Hydrometallurgy 156:124–135
9. Liu XH, Chen DF, Li YK et al (2013) Investigation on efficient vanadium extraction from vanadium bearing molten iron. Adv Mater Res 785–786:58–62
10. Peng Y, Xie T, Zhou Z, Pan P (2007) Preparation V_2O_5 from low grade vanadium-bearing slag of high calcium and high phosphor. Ferro-Alloy 195(4):18–23
11. Zhang H (2003) Determination of the optimum conditions of roasting of vanadium slag by orthogonal test. Inorg Chem Ind 35(3):29–31
12. Jena BC, Dresler Reilly IG (1995) Extraction of titanium, vanadium and iron from titanomagnetite deposits at Pipestone Lake. Miner Eng 8(1–2):159–168

Part II
Rare Metals II

Supercritical Fluid Extraction for Urban Mining of Rare Earth Elements

Jiakai Zhang, John Anawati, Yuxiang Yao and Gisele Azimi

Abstract For moving toward a sustainable future and building the circular economy, there is a push toward waste valorization and urban mining. NiMH batteries and NdFeB magnets contain significant amounts of rare earth elements (REEs). These elements offer unique physicochemical properties; thus, they have become a key component of many critical green technologies and some are identified as critical material and initiatives have begun to recycle them from postconsumer products. Conventional recycling processes rely on pyrometallurgy or hydrometallurgy. The former is energy intensive, generating greenhouse emissions, while the latter consumes large volumes of acids and organic solvents and generates vast amounts of hazardous waste. Here, we develop an environmentally sustainable recycling process, in which we utilize supercritical fluids to extract REEs from end-of-life products. We believe that supercritical fluid extraction (SCFE) process is a viable alternative to pyrometallurgy and hydrometallurgy for urban mining of REEs in an economical and environmentally benign manner.

Keywords Rare earth elements · Supercritical fluids extraction · Recycling
Urban mining · NdFeB magnet

Jiakai Zhang, John Anawati, Yuxiang Yao—These authors contributed equally to this work.

J. Zhang · J. Anawati · Y. Yao · G. Azimi (✉)
Laboratory for Strategic Materials, Department of Chemical Engineering and Applied Chemistry, 200 College Street, Toronto, ON M5S 3E5, Canada
e-mail: g.azimi@utoronto.ca

G. Azimi
Department of Materials Science and Engineering, University of Toronto, 184 College Street, Toronto, ON M5S 3E4, Canada

G. Azimi et al. (eds.), *Rare Metal Technology 2019*, The Minerals, Metals & Materials Series, https://doi.org/10.1007/978-3-030-05740-4_7

Introduction

With increasing awareness of environmental protection, critical green technologies, including wind turbines, electric vehicles, photovoltaic cells, and fluorescent lighting, among others, are widely promoted and developed. Many of these green technology developments and production depend on the unique physiochemical properties provided by rare earth elements (REEs) [1]. Some REEs like neodymium (Nd), praseodymium (Pr), and dysprosium (Dy) are presently regarded as critical elements according to both the US Department of Energy and the European Commission [2, 3].

Neodymium-iron-boron (NdFeB) magnets are a class of permanent magnets that are in increasing demand because of their unique application in wind turbines and electric vehicles. It has been reported that the production of electric vehicles is forecasted to increase from 2.3 million units in 2016 to over 10.1 million units in 2026 [4]. Similarly, global wind energy council has reported a 12.6% cumulative capacity growth rate reaching a total of 486.6 GW in 2016, and they forecasted that the cumulative installed capacity will reach over 800 GW by the end of 2021 [5]. As the utilization rate of NdFeB magnets increases in the green technologies, there is an urgent need for managing end-of-life magnets in a cost-effective, efficient, and environmentally sustainable manner. Such an approach would result in the urban mining of valuable elements, including Nd, Pr, and Nd.

Commonly available recycling processes rely on pyrometallurgy or hydrometallurgy. Pyrometallurgy is energy intensive and generates large volume of greenhouse gas emission. Hydrometallurgy is the preferred process; however, it faces several drawbacks, including consumption of large volumes of reagents, and generation of large volumes of hazardous waste.

Supercritical fluid extraction (SCFE) is an emerging green separation technology that has been used to recover metals from postconsumer products [1]. It has been developed in late 1970s to extract specialty products and then expanded to less specialized applications. Most commercial SCFE processes have focused on energy industry, including direct liquefaction of coal [6] and enhanced oil recovery from petroleum reservoirs [7], and on chemical industry, including food and pharmaceuticals. A supercritical fluid is defined as a substance at a temperature and pressure beyond its critical points, where there are no distinct liquid and gas phases. It can effuse through solids like a gas and dissolve materials like a liquid. Unique properties of supercritical fluids are high diffusivity and solvability, and low viscosity. Those properties enable them to penetrate and transport solutes in different matrices at a higher rate and more efficient than that in a liquid phase.

Close to the critical point, small changes in pressure or temperature result in large changes in density, allowing many properties of a supercritical fluid to be fine-tuned. Carbon dioxide and water are the most commonly used supercritical fluids, being utilized for decaffeination and power generation, respectively. Carbon dioxide has moderate critical properties ($Tc = 31.1$ °C, $Pc = 7.38$ MPa, $\rho = 471$ kg/m^3). Furthermore, it is inert, safe, abundant, and easy to vent or recycle. Supercritical CO_2 is a non-polar solvent; thus, there is a significant polarity difference between the solute and solvent when extracting metal ions and organometallic compounds. Therefore, it is necessary to utilize complexing agents to provide the charge neutrality and improve solvent–solute interactions.

In 2017, Yao et al [1] developed a supercritical fluid extraction process to recycle REEs from postconsumer NiMH battery. Supercritical CO_2 with TBP-HNO_3 chelating agent was utilized, and they achieved 86% La, 86% Ce, 88% Pr, and 90% Nd recovery. The effect of seven operating parameters, including temperature, pressure, residence time, sample-to-chelating agent ratio, agitation rate, complex chemistry, and cosolvent (methanol) addition, on the REE extraction efficiency was studied. The optimum conditions were reported at 35 °C, 31 MPa, 2 h residence time, S:CA ratio of 1:10, and $TBP(HNO_3)_{1.745}(H_2O)_{0.52}$ chelating agent with 2 mol% methanol addition.

In this study, SCFE process is developed to extract REEs from end-of-life NdFeB magnets. A fractional factorial design of experiment was utilized to design the experimental matrix and to obtain the optimum operating conditions in terms of pressure, temperature, solid-to-chelating agent ratio (S:CA), residence time, and agitation rate with and without cosolvent (methanol) addition, to determine optimal operating conditions.

Experimental Design

The process flow diagram developed in this study is presented in Fig. 1. The entire system including the 100-mL high-pressure rated reactor, magnetic drive mixer, reactor controller, constant flow dual piston pump, and solvent pump was manufactured by Supercritical Fluid Technology Inc., USA. A known amount of magnet material and TBP-HNO_3 was loaded into the reactor. The reactor head was closed, and liquid CO_2 was pumped into the chamber with both restrictor valve and static/dynamic valve closed until the desired pressure was reached. The reactor was heated using the electric heating jacket surrounding the reactor chamber. During the reaction, soluble rare earth complex is formed; after completing the process, it was collected in the EPA vial by opening both restrictor valve and static/dynamic valve. Reproducibility tests (three independent experiments) showed that the experimentally measured data are accurate to within $\pm 5\%$. In this process, leaching efficiency was calculated as follows:

$$LeachingEfficiency, \% = \left(1 - \frac{w_r}{w_o}\right) * 100\% \quad (1)$$

where w_r is the weight of residue and w_o is the original weight of that meal in magnet sample.

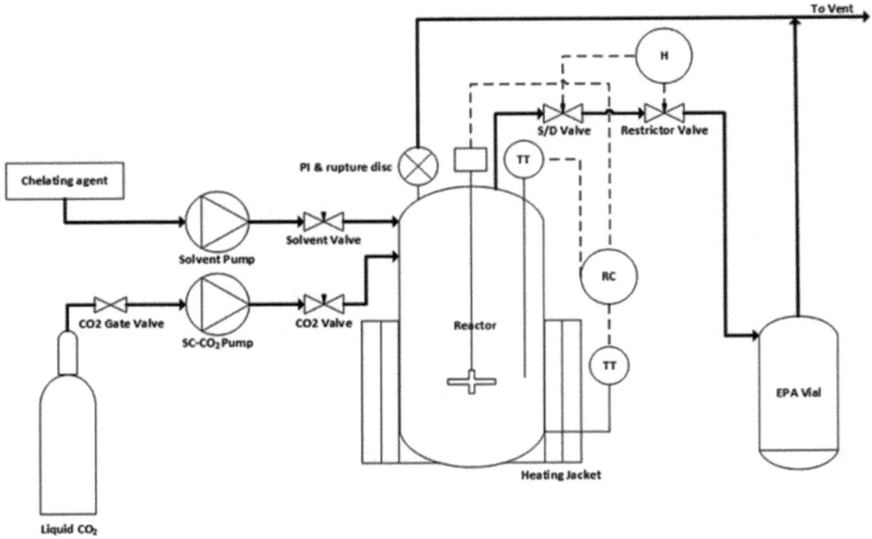

Fig. 1 SCFE process flow diagram. Adopted from Yao et al. [1]

Fractional Factorial Design of Experiment Methodology

To work with multi-variables, the traditional experimental design method, such as one factor at a time, is inefficient and tedious. Also, zero information will be produced on interaction impacts due to two or more variables. Hence, a fractional factorial design of experiment methodology is utilized, with the goal of testing a large number of processing parameters, while maximizing workflow efficiency by reducing the total required number of trials.

The parameters selected for testing were as follows: the extraction temperature (X_1), the extraction pressure (X_2), the extraction duration (X_3), the sample: chelating agent ratio (X_4), the agitation rate (X_5), the HNO_3 content of the chelating agent (X_6), and the use of methanol as a cosolvent (X_7). The test levels were selected on the basis of the levels used in Yao et al.'s study on the supercritical CO_2 extraction of REEs from NiMH batteries [1]. The same levels were used for both studies to allow direct comparison between the REE battery system and the REE magnets system. The parameter settings (X_i) were normalized and coded between −1 (low level) and +1 (high level), according to Eq. 2—the parameter coding allows direct comparison of the relative impact of each of the experimental parameters on the system response by direct comparison of the magnitude of the model coefficients. A summary of the levels associated with each of the test parameters is given in Table 1.

$$X_i = \frac{x_i - \frac{1}{2}(x_{\max} + x_{\min})}{\frac{1}{2}(x_{\max} - x_{\min})} \quad (2)$$

Table 1 Summary of test parameters and associated levels

Factor	Units	−1 level	0 level	+1 level
X_1: temperature	°C	35	45	55
X_2: pressure	MPa	20.7	25.85	31
X_3: duration	h	1	1.5	2
X_4: S:CA ratio	g/ml	0.20	0.15	0.10
X_5: agitation	rpm	750	1125	1500
X_6: chelating agent	M	10.4	13.05	15.7
X_7: methanol addition	%	0	1	2

Results and Discussion

Characterization of Ground NdFeB Magnet Powder

A comprehensive characterization of the elemental composition, mineralogical characteristics, microstructure, and macrostructure of the ground NdFeB magnets was performed. For use in the computation of extraction efficiency, the elemental composition of the magnets in iron and REEs (neodymium, praseodymium, and dysprosium) was determined by microwave digesting the crushed magnet powder in aqua regia, and quantification by inductively coupled plasma optical emission spectroscopy (ICP-OES). The elemental composition is presented in Fig. 2. The magnets comprise primarily Fe, Nd, and Pr, with an Fe:REE ratio of approximately 2.3:1. This is slightly less than the 2.7:1 ratio suggested by the stoichiometry of tetragonal $Nd_2Fe_{14}B$, suggesting that these magnets are enriched in REEs compared with the expected alloy structure. As shown in X-ray diffraction (XRD) spectrum in Fig. 3, the major phase in the magnet sample is $Nd_2Fe_{14}B$, with a small amount of $Nd(OH)_3$. Figure 4 presents a backscattered secondary electron (BSE) image of the magnet particle. As can be seen, dark and bright spots were detected, and they were evenly distributed across the magnet. The bright spots normally indicate the presence of heavy elements, in this case neodymium, so it is mainly composed of $Nd(OH)_3$. The dark regions represent $Nd_2Fe_{14}B$. The particle size distribution of the fine magnet particles is shown in Fig. 5, indicating average particle size of 21.6 μm.

Extraction Results

On the basis of previous experimental experience and a thorough review of previously reported supercritical extraction experiments, a range of process parameters and appropriate parameter levels were selected during the design experiment (Table 1). The relative effect of each of these factors was assessed in a saturated 2^{7-4}_{III} fractional factorial test, including three replicate centre points. The resolution of this

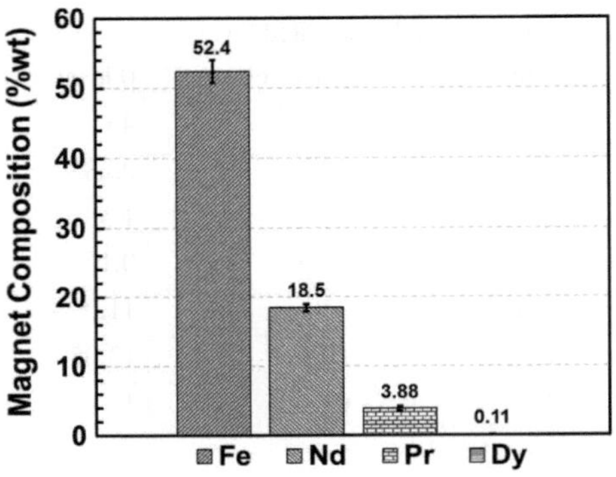

Fig. 2 Ground NdFeB magnet powder characterization results. ICP-OES elemental analysis, error bars represent the sample standard deviation (nFe, Nd, Dy = 5, nPr = 2)

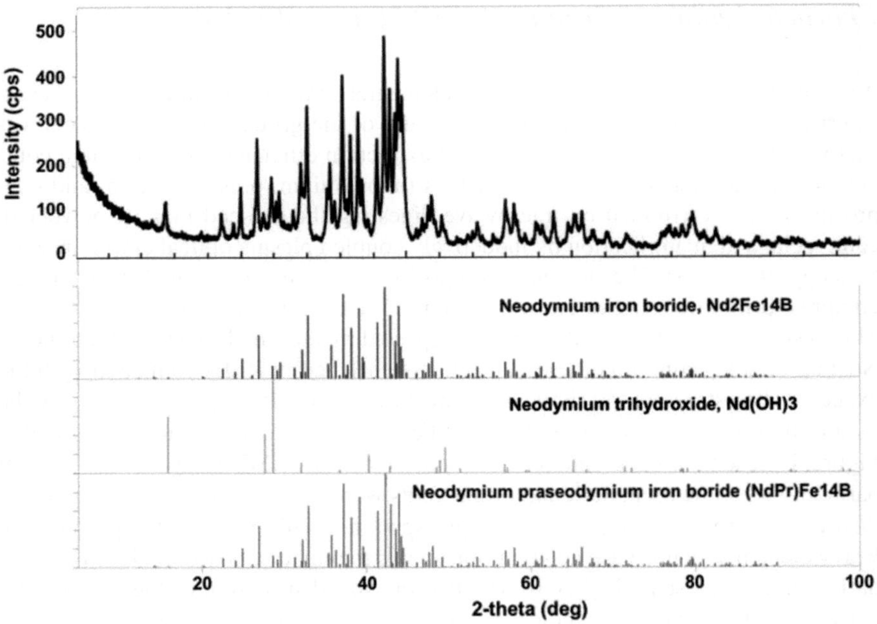

Fig. 3 X-ray diffraction spectrum with the identified crystalline phases

test was then increased from *III* to *IV* with a mirror-image fold-over test, allowing for unaliased estimation of the primary test factors. The accuracy of the empirical model constructed from these tests was then verified by a series of validation tests which allow an independent comparison of the experimental extraction results and the model-predicted results. The extraction results for the 19 runs are presented in Fig. 6.

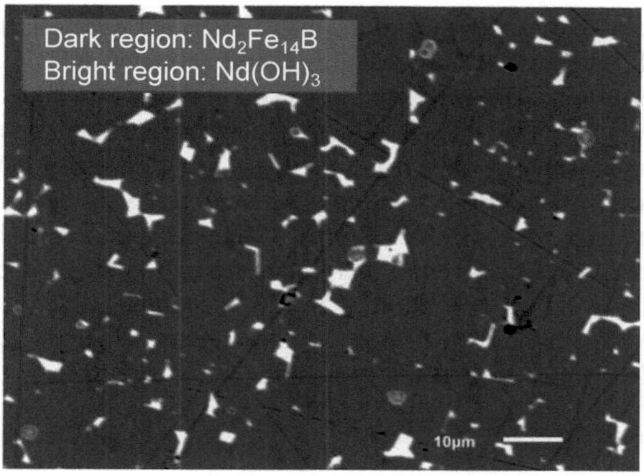

Fig. 4 Scanning electron micrograph (backscattered) of the magnet surface

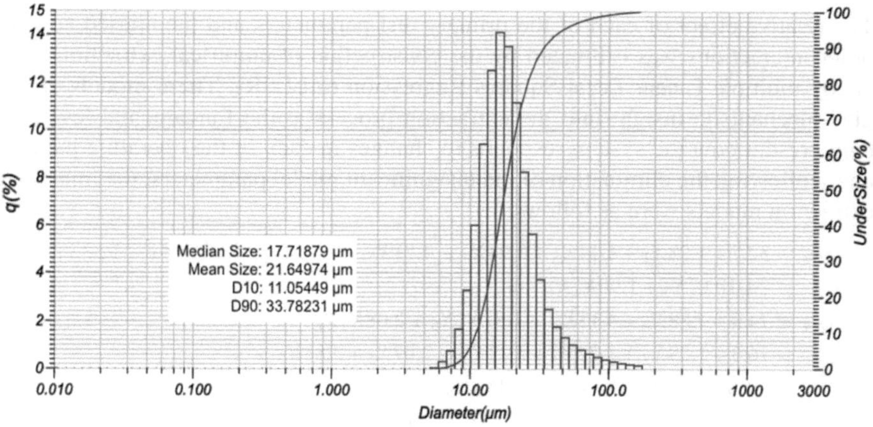

Fig. 5 Particle size distribution

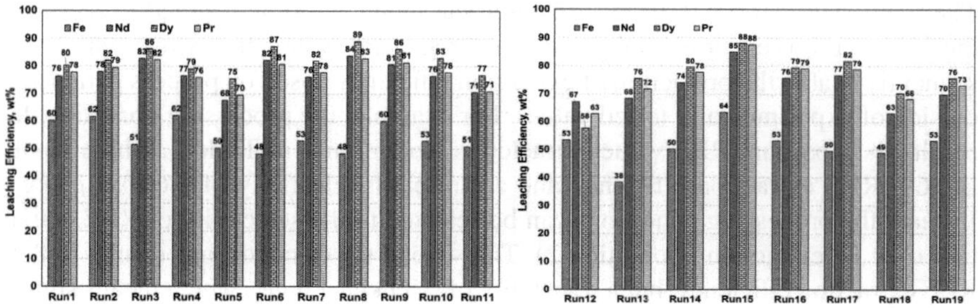

Fig. 6 Extraction efficiency of Fe, Nd, Dy, and Pr for the entire experimental matrix

The parameter settings for each of the tests performed were selected by the fractional factorial design algorithm; because several variables are adjusted at once in each trial, the effect of each main experimental parameter was assessed by comparing the average extraction responses of all tests to each of these parameters, as opposed to pairwise comparison of results from individual tests. Fe extraction was generally considerably lower than REE extraction, with an average extraction of 53%, compared with an average of 77% for the REEs. The factors appearing to have the largest positive impact on REE extraction were an increased amount of chelating agent (X_4) and an increased concentration of HNO_3 in the chelating agent (X_6). These results were used to build empirical extraction models, which allow a quantitative comparison of the relative impact of each of the tested parameters.

Empirical Extraction Modeling

The relative effect of each of the primary test parameters and the aliased second-order interactions was estimated by the construction of empirical extraction models by multiple Linear Least Squares Regression (mLLSR). These models do not give mechanistic insight into the extraction process at the fundamental level, but they present how the system responds to changes in each of the operating parameters. Following the elimination of the insignificant model parameters, the final fitted extraction models were as follows:

$$\begin{aligned}
\hat{y}_{Fe} &= 52.7 + 1.08X_1 + 1.14X_3 + 0.94X_4 - 4.71X_5 - 1.88X_7 - 1.58X_2X_4 - 1.53X_1X_2 \\
&\quad - 1.10X_1X_3 + 1.90X_1X_6 \\
\hat{y}_{Nd} &= 74.9 + 2.88X_4 + 4.13X_6 - 1.81X_1X_5 + 2.08X_1X_6 \\
\hat{y}_{Pr} &= 76.3 + 1.29X_2 + 1.72X_3 + 2.13X_4 - 2.07X_5 + 3.88X_6 \\
&\quad - 1.36X_7 - 1.77X_1X_4 - 2.08X_1X_5 + 2.71X_1X_6 \\
\hat{y}_{Dy} &= 79.4 + 1.28X_3 + 2.05X_4 - 1.17X_5 + 4.13X_6 \\
&\quad - 1.66X_4 - 1.91X_1X_5 + 1.29X_1X_3 + 2.22X_1X_6
\end{aligned}$$

The fractional factorial tests and the empirical model building resulted in accessing the effect various experimental parameters on general system behaviour, which subsequently enable the prediction of extraction efficiency based on process parameters outside of experimentally tested values. For example, the processing condition that results in maximum REE value extracted is Condition 1 in Table 2, which results in 86% REE extraction efficiency and a rare earth selectivity of 0.60 kg_{REE}/kg_{Fe}. Instead, the processing conditions can be selected to the selectivity of REE extraction over Fe extraction (Condition 2). The second set of conditions results in 82% REE extraction efficiency, but a selectivity of 0.89 kg_{REE}/kg_{Fe}, which means significantly higher REE enrichment, together with other economically positive conditions such as low process pressure and duration. Therefore, Condition 2 was selected as the optimum condition in this study.

Table 2 Optimized process parameters and predicted outputs for two design cases: maximum REE value recovery and maximum REE selectivity. (S:CA = Solids:Chelating Agent ratio)

Temperature X_1 (°C)	Pressure X_2 (MPa)	Time X_3 (h)	S:CA X_4 (g/mL)	Agitation X_5 (rpm)	HNO_3 X_6 (M)	MeOH X_7 ($\%_{mol}$)	Extraction (%) Nd, Dy, Pr, Fe	Value recovery (%)	Selectivity (kg_{REE}/kg_{Fe})
Condition 1: maximum REE value recovery									
55	25.8	1.5	0.1	750	15.7	1	85.8, 88.8, 85.9, 60.4	86	0.60
Condition 2: maximum REE selectivity									
55	20.7	1	0.1	1500	15.7	2	82.3, 88.8, 77.9, 38.6	82	0.89

Conclusions

An environmentally friendly supercritical fluid extraction process was developed to extract REEs from end-of-life NdFeB magnet. The effect of seven operating parameters, namely temperature, pressure, residence time, sample-to-chelating agent ratio, agitation rate, complex chemistry, and cosolvent (methanol) addition, on the REE extraction efficiency was investigated, and 55 °C, 21 MPa, 1 h residence time, S:CA ratio of 1:10, $TBP(HNO_3)_{1.745}(H_2O)_{0.52}$ chelating agent with 2 mol% methanol addition were determined to be the optimum conditions, resulting in 82% Nd, 89% Dy, 78% Pr, and 39% Fe extraction. The developed process requires low content of chelating agent and generates minimal hazardous waste. Moreover, energy consumption of this process is significantly lower than that of a pyrometallurgical process.

References

1. Yao Y, Farac NF, Azimi G (2018) Supercritical fluid extraction of rare earth elements from nickel metal hydride battery. ACS Sustain Chem Eng 6:1417–1426
2. Chu S (2011) Critical Materials Strategy 2011. U.S. Department of Energy. DOE/PI-0009
3. Golev A, Scott M, Erskine PD, Ali SH, Ballantyne GR (2014) Rare earths supply chains: current status, constraints and opportunities. Resour Policy 41:52–59
4. Goodenough KM, Wall F, Merriman D (2017) The rare earth elements: demand, global resources, and challenges for resourcing future generations. Nat Resour Res 27:201–216
5. Global Wind Energy Council (2016) Global wind report 2016—annual market update. Global Wind Report 2016
6. Kershaw JR, Jezko J (1982) Supercritical gas extraction of South African coals. Sep Sci Technol 17:151–166
7. Ely JF, Baker JK (1983) A review of supercritical fluid extraction. Government Printing Office, U.S

Recovery of Scandium by Leaching Process from Brazilian Red Mud

Amilton Barbosa Botelho Junior, Raquel Húngaro Costa, Denise Crocce Romano Espinosa and Jorge Alberto Soares Tenório

Abstract Alumina is the raw material for aluminum production, where it is obtained from bauxite by the Bayer Process. Sodium hydroxide is used in a pressure digestion with to obtain an aluminate solution and then remove impurities by the precipitation. For each 1 ton of alumina produced, 1–1.25 tons of red mud is generated in the Bayer Process. Red mud is mainly composed of iron, silicon and titanium oxides, but rare earth elements (REEs) can also be found in the residue. The most important REE is scandium, because it is 90% of the price of all REEs present in red mud. The aim of this work was to study a leaching process of Brazilian red mud to obtain scandium. Sulfuric acid 20 wt% was used as leaching agent, L/S ratio 1/10 (please add unit), temperature of 25 °C and 6 h of reaction time. Samples were analyzed using inductively coupled plasma optical emission spectroscopy (ICP-OES) and X-ray diffraction (XRD).

Keywords Scandium · Rare earth elements · Bauxite · Bayer process

Introduction

The Bayer process is the main process applied to obtain alumina from bauxite ore. The mineral is dissolved using sodium hydroxide in digesters, which obtain sodium aluminate in solution and a solid phase called red mud; the two are separated by the clarification process [1, 2]. Brazil is the third country in bauxite reserves in the world, while Guinea and Australia are the first and the second, respectively. Brazil is the fourth in bauxite production, while Australia, China and Guinea are the first, the second and the third in the world [3].

A. B. Botelho Junior (✉) · R. H. Costa · D. C. R. Espinosa · J. A. S. Tenório
Department of Chemical Engineering, Polytechnic School of
University of Sao Paulo, Sao Paulo, Brazil
e-mail: amilton.junior@usp.br

G. Azimi et al. (eds.), *Rare Metal Technology 2019*, The Minerals, Metals & Materials Series, https://doi.org/10.1007/978-3-030-05740-4_8

The main problem of alumina production by the Bayer process is red mud generation. In 2000, the alumina industry had produced about 2 million tons of red mud, and it was estimated to reach 4 billion tons in 2015 [4]. Different applications have been studied for red mud, geopolymers, as catalysis, composite materials, and a secondary source to recover valuable metals [5].

Red mud around the world has been studied for REE recovery. China is the most important producer of REEs from mines; however, due to Chinese export restrictions since 2011, alternatives to obtain REEs have been studied [6–8], and bauxites are the most promising resources of REE, mainly scandium. Scandium represents 90–95% of the economic value of the REE present in red mud [9]. Borra et al. studied the alumina and iron recovery by alkali roasting and smelting process, and a leaching process to recover REEs [10].

The goal of this work was to study the recovery of scandium from Brazilian red mud (BRM) by leaching process. Sulphuric acid 20 wt% was used as leaching agent, S/L ratio was set at 1/10 and reaction time was set at 6 h. The reaction temperature was controlled at 25 °C. Samples of liquid phase were analyzed using ICP-OES, and the solid phase was analyzed using scanning electron microscopy energy dispersive spectroscopy (SEM/EDS) and XRD.

Materials and Methods

Brazilian red mud (BRM) sample was dried at 60 °C for 24 h. Sulfuric acid 20 wt% P.A. was prepared with deionized water. Leaching experiment was performed at 25 °C, for 6 h in a stirring speed of 800 rpm. Solid/Liquid ratio was 1/10. A sample of 50 g of BRM was added in a glass-reactor Atlas Sodium 1.4 (Syrris) with sulfuric acid solution (Fig. 1).

Table 1 presents the elemental composition of BRM (%). A sample of residue was digested in a microwave Mars 6 with 4 different acids (sulfuric, hydrochloride,

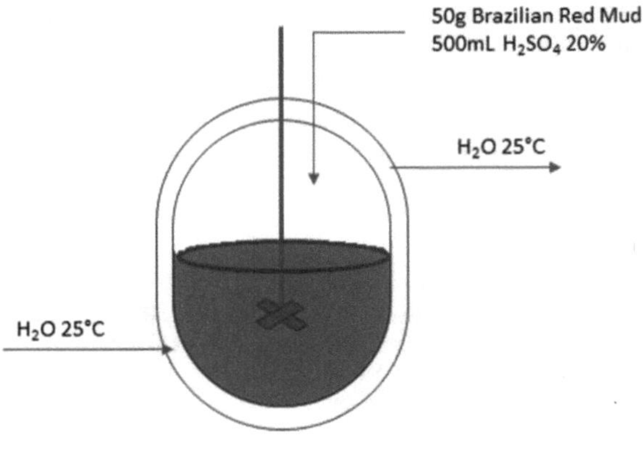

Fig. 1 Illustration of the leaching experiment in a glass-reactor with controlled temperature

Table 1 Elemental composition of Brazilian red mud, in wt%

Elements	wt%
Fe	18.94
Al	9.18
Si	7.52
Ca	3.11
Na	0.336
Ti	1.416
Nb	0.136
Sc	0.0043
Y	0.0024
Zr	0.132
Others	59.22

nitric and hydrofluoric) and analyzed using ICP-OES. Iron is the main metal in the residue—18.94 wt%—followed by aluminum and silicon—9.18 and 7.52 wt%, respectively. Niobium, yttrium, zirconium and scandium are also present at lower concentration. Other elements (56.22 wt%) should be organic compounds [11, 12] or elements that were not analyzed, such as oxygen and hydrogen. Among all metals in the BRM, iron is the main element (50.74 wt%), while scandium concentration is 43 mg/kg in BRM. After the leaching process, samples of liquor were analyzed using ICP-OES. The solid phase obtained was analyzed using XRD and SEM-EDS.

Results and Discussion

Sulfuric acid 20 wt% was first added to the glass-reactor and the BRM sample was added after temperature reached equilibrium (25 °C). Leaching process performed for 6 h, and then the solution was filtrated. Liquor samples obtained were analyzed using ICP-OES, and percentage of metals recovered is presented in Table 2. Sodium and aluminum were the elements with higher percentage leached (44.23 and 22.38%, respectively). Among elements in lower concentration, scandium and yttrium recovery percentage were close (8.96 and 8.81%, respectively), which indicates that both metals recovery is similar, although their concentrations are different. Iron and silicon are the elements in higher concentration (first and third) in BRM, and the irrecovery was 0.46 and 0.05%, respectively.

Table 2 Percentage of metals leached using sulfuric acid 20 wt%, S/L ratio 1/10, at 25 °C during 6 h

Element	Fe	Al	Si	Ca	Na	Ti	Nb	Sc	Y	Zr
%	1.02	20.44	0.14	2.96	44.85	1.87	3.29	9.65	9.43	13.99

Reid et al. studied the leaching process of Canadian red mud with sulphuric acid, where it was possible to recover ~28% of scandium at 25 °C with 3M of sulphuric acid at a S/L of 1/15 for 30 min [13]. Results presented in Table 2 were obtained using 3.75M of sulphuric acid (20%) and S/L of 1/10, which indicates that acid concentration and S/L ratio have an effect on scandium recovery.

In the concentration of 20 wt%, solution of sulfuric acid has a pH of −0.88, calculated using Eqs. 1 and 2. After the leaching process, the pH of the liquor obtained was −0.35, which indicates that 2.59 mol/L of sulfuric acid was consumed (1.29 mol, or 126.67 g). Acid consumption value is important for economic aspects, to study the effect of acid concentration and also to understand if it is necessary to use excess sulfuric acid in the leaching process.

$$H_2SO_4 \rightarrow 2H^+ + SO_4^{-2} \tag{1}$$

$$\mathrm{M} = \frac{\mathrm{m}}{\mathrm{MM.V_T}} \tag{2}$$

m is the concentration (mol/L), MM is the molar mass (g/mol) and V_T is the total volume (500 mL).

Figure 2 presents the SEM image (5000x) of (a) crude sample of BRM and (b) the solid phase filtrated after the leaching process (c) EDS spectra. The SEM images show the effect of sulphuric acid on the leaching process of BRM, where the particles were agglutinated. The EDS spectra present the major elements present in the solid phase, such as titanium, oxygen, iron, silicon, aluminum, and sulphur. Sulphates and oxides could be formed due to the presence of oxygen and sulphur.

Figure 3 presents the X-ray diffractogram of the solid phase filtrated after the leaching process. The numbers indicate the phases identified in XRD, presented in Table 3. The phases identified were rhomboclase, tamarugite, aluminum oxide, quartz, cristobalite, hematite and magnetite. The sulphates formed were rhomboclase ($HFe(SO_4)_2{\cdot}4(H_2O)$) and tamarugite ($NaAl(SO_4)_2{\cdot}6H_2O$). The oxides formed were aluminum oxide (Al_2O_3), quartz and cristobalite (SiO_2), hematite (Fe_2O_3) and magnetite (Fe_3O_4).

Alkan et al. studied the leaching process with sulphuric acid and hydrochloride acid. The authors found, in the solid phase obtained after leaching with sulphuric acid, iron oxide (goethite—FeO), rhomboclase, aluminum oxide (diaspore—α-AlO(OH), and boehmite—γ-AlO(OH)), and silicon oxide (quartz and critobalite—SiO_2). Gypsum ($CaSO_4{\cdot}2H_2O$), titanium oxide and other phase content with aluminum and silicon were also found [14].

In another study, during leaching of scandium and titanium using sulphuric acid and hydrogen peroxide, boehmite, rhomboclase, diaspore, hematite and quartz were found [15]. Which indicates the similarity with the phases found after the leaching process is due to the similarity in all red mud composition, in terms of iron, aluminum, silicon and others elements.

Fig. 2 SEM images of **a** crude sample, **b** the solid phase filtrated after the leaching process and **c** EDS spectra

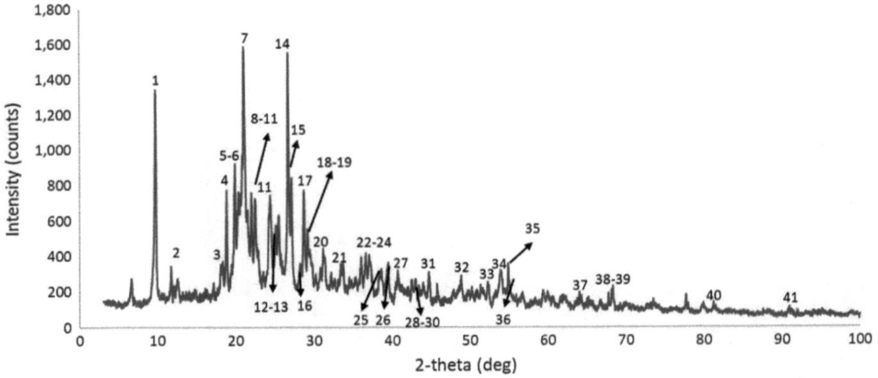

Fig. 3 X-ray diffractogram of residue filtrated after leaching process

Table 3 Compounds detected in XRD of residue filtrated after the leaching process

Phase	Formula	No.
Rhomboclase	$HFe(SO_4)_2{\cdot}4(H_2O)$	1, 3, 4, 6, 7, 9, 11, 12, 14, 15, 17, 19, 20, 21, 24, 25, 26, 27, 29, 30, 31, 32, 33, 34, 36, 37, 38, 39
Tamarugite	$(NaAl(SO_4)_2{\cdot}6H_2O)$	2, 3, 6, 8, 10, 11, 12, 15, 16, 17, 18, 19, 20, 21, 22, 23, 24, 25, 26, 27, 28, 30, 31, 32, 33, 34, 35
Aluminum oxide	Al_2O_3	5, 13, 25, 27, 30, 34, 37, 40, 41
Quartz	SiO_2	7, 14, 24, 26, 27, 35, 27, 40, 41
Cristobalite	SiO_2	9, 20, 31, 34, 40
Hematite	Fe_2O_3	13, 21, 25, 30, 33, 40, 41
Magnetita	Fe_3O_4	25, 30, 39, 40, 41

Conclusion

The goal of this work was to study the leaching process of Brazilian red mud using sulphuric acid for scandium recovery. Scandium recovery was 9.65%, using sulphuric acid 20 wt%, S/L ratio 1:10, during 6 h in a stirring speed of 800 rpm at 25 °C. Iron recovery and aluminum recovery were 1 and 20%, respectively. The phases found in the filtrated were also reported in literature, probably due to the similarity in the red mud composition. Futures studies must focus on the leaching process to obtain scandium in an economic route.

Acknowledgements To University of Sao Paulo for support of this project. To Sao Paulo Research Foundation and Capes (FAPESP—grant: 2018/11417-3, 2018/03483-6) for financial support.

References

1. Huo SH, Qian M, Schaffer GB, Crossin E (2011) Fundamentals of aluminium metallurgy, 1st edn. Woodhead Publishing, Cambridge. https://doi.org/10.1533/9780857090256.3.655
2. Smith P (2009) The processing of high silica bauxites—review of existing and potential processes. Hydrometallurgy 98:162–176. https://doi.org/10.1016/j.hydromet.2009.04.015
3. Plunkert P (2017) Bauxite and alumina, US Geol Surv Miner Yearb 1 26–27. http://scholar.google.com/scholar?hl=en&btnG=Search&q=intitle:Bauxite+and+alumina+1#6
4. Wang W, Pranolo Y, Cheng CY (2013) Recovery of scandium from synthetic red mud leach solutions by solvent extraction with D2EHPA. Sep Purif Technol 108:96–102. https://doi.org/10.1016/j.seppur.2013.02.001
5. Verma AS, Suri NM, Kant S (2017) Applications of bauxite residue: a mini-review. Waste Manag Res 35:999–1012. https://doi.org/10.1177/0734242X17720290
6. Narayanan RP, Kazantzis NK, Emmert MH (2018) Selective process steps for the recovery of scandium from Jamaican bauxite residue (red mud). ACS Sustain Chem Eng 6:1478–1488. https://doi.org/10.1021/acssuschemeng.7b03968
7. Chen Z (2011) Global rare earth resources and scenarios of future rare earth industry. J Rare Earths 29:1–6. https://doi.org/10.1016/S1002-0721(10)60401-2
8. Walawalkar M, Nichol CK, Azimi G (2016) Process investigation of the acid leaching of rare earth elements from phosphogypsum using HCl, HNO_3, and H_2SO_4. Hydrometallurgy 166:195–204. https://doi.org/10.1016/j.hydromet.2016.06.008
9. Rivera RM, Ulenaers B, Ounoughene G, Binnemans K, Van Gerven T (2018) Extraction of rare earths from bauxite residue (red mud) by dry digestion followed by water leaching. Miner Eng 119:82–92. https://doi.org/10.1016/j.mineng.2018.01.023
10. Borra CR, Blanpain B, Pontikes Y, Binnemans K, Van Gerven T (2017) Recovery of rare earths and major metals from bauxite residue (red mud) by alkali roasting, smelting, and leaching. J Sustain Metall 3:393–404. https://doi.org/10.1007/s40831-016-0103-3
11. Power G, Loh J (2010) Organic compounds in the processing of lateritic bauxites to alumina Part 1: origins and chemistry of organics in the Bayer process. Hydrometallurgy 105:1–29. https://doi.org/10.1016/j.hydromet.2010.07.006
12. Power G, Loh JSC, Vernon C (2012) Organic compounds in the processing of lateritic bauxites to alumina Part 2: effects of organics in the Bayer process. Hydrometallurgy 127–128:125–149. https://doi.org/10.1016/j.hydromet.2012.07.010
13. Reid S, Tam J, Yang M, Azimi G (2017) Technospheric mining of rare earth elements from bauxite residue (red mud): process optimization, kinetic investigation, and microwave pretreatment. Sci. Rep. 7:1–9. https://doi.org/10.1038/s41598-017-15457-8
14. Alkan G, Schier C, Gronen L, Stopic S, Friedrich B (2017) A mineralogical assessment on residues after acidic leaching of bauxite residue (red mud) for titanium recovery. Metals (Basel) 7:458. https://doi.org/10.3390/met7110458
15. Alkan G, Yagmurlu B, Cakmakoglu S, Hertel T, Kaya Ş, Gronen L, Stopic S, Friedrich B (2018) Novel approach for enhanced scandium and titanium leaching efficiency from bauxite residue with suppressed silica gel formation. Sci Rep 8:5676. https://doi.org/10.1038/s41598-018-24077-9

Selective Precipitation of Th and Rare-Earth Elements from HCl Leach Liquor

Haydar Güneş, Hüseyin Eren Obuz and Murat Alkan

Abstract In this study, selective precipitation of thorium and rare-earth elements was investigated. The leach liquor contains Th, Ce, La, Nd, Pr, and Sm. Precipitation process was realized by using oxalic acid and selectivity of process was provided with the addition of oxalic acid at different volumes. The concentration of oxalic acid and settling duration were selected as parameters. The pH change of leach liquor was controlled during the experiments. The precipitation order was identified as Th, Sm, Nd and Pr, Ce, and La due to increasing of the oxalic acid volume. The precipitation efficiencies were calculated with remaining metals in leach liquor. The elemental analysis of solutions was realized with Inductively Coupled Plasma—Optical Emission Spectrometer (ICP—OES), and the phase analysis of the precipitates was realized with X-ray diffractometer (XRD).

Keywords Precipitation · Rare-earths · Thorium

Introduction

The rare-earth elements (REEs) are a group of metallic elements that share similar chemical properties. The lanthanides, scandium, and yttrium were defined as the REE. According to their similarity, they exist together in minerals and separating them to create a problem for scientific and industrial applications [1]. World trends were changed into green energy, and the REEs have important roles in green energy applications. Due to this change, the demand for REEs was increased [2]. The REEs can be found in more than 250 minerals, but only 200 of them have more than 0.01%

H. Güneş · H. E. Obuz · M. Alkan (✉)
Department of Metallurgical and Materials Engineering, Faculty of Engineering, Dokuz Eylül University, 35397 Buca, Izmir, Turkey
e-mail: alkan.murat@deu.edu.tr

H. Güneş · H. E. Obuz
The Graduate School of Natural and Applied Sciences, Dokuz Eylül University, 35397 Buca, Izmir, Turkey

G. Azimi et al. (eds.), *Rare Metal Technology 2019*, The Minerals, Metals & Materials Series, https://doi.org/10.1007/978-3-030-05740-4_9

REEs. The most important minerals for the industry were bastnaesite, monazite, xenotime, and parasite. These four minerals have 95% of the world REEs reserve [1, 3]. Bastnaesite and monazite are the primary sources for the light REEs (Ce, La, Nd, Pr, and Sm). Xenotime is the main source for the heavier REEs. Y is the principal REE and Gd, Dy, Ho, Er, and Yb are the other REEs in the xenotime. Monazite contains Th up to 20 wt%, and bastnaesite also contains Th in small amounts [4].

The traditional REE production routes consist of ore preparation, roasting, leaching, solvent extraction, and precipitation. Solvent extraction process has three stages: extraction, scrubbing, and stripping. These stages were repeated several times for separating REEs [5].

Some researchers were tried to precipitate REEs after leaching but in these studies, they were focused on leaching stage and precipitation of REEs all together. After this process, a REE's mixture can be obtained from the sulfate solution as oxalates on the basis of Eq. (1) [6]. A general reaction can be written as below: Eq. (2).

$$RE_2(SO_4)_3 + 3H_2C_2O_4 \rightarrow RE_2(C_2O_4)_3 + 3H_2SO_4 \qquad (1)$$

$$RE_xA_y + yH_2C_2O_4 = RE_x(C_2O_4)_y + yH_xA \qquad (2)$$

Thorium can be separated from REEs in the solvent extraction process. Precipitation of Th and REEs was generally applied after stripping from organic solutions. For precipitating most of the REEs that obtained from Mongolian Apatite Ore by H_2SO_4, oxalic acid/total REE's molar ratio determined as 0.3 [7]. REE affinity for oxalic acid is high if Th is not in solution, but REEs are precipitated together with oxalic acid. The RE-oxalate compounds can be calcined to obtain RE-oxides. RE-oxides are an important commercial compound for REEs [8].

The aim of this study is investigating the selective precipitation potential of the Th and complex REE's leach liquor with oxalic acid. After addition of the oxalic acid, solutions were filtered and ICP–OES analysis was realized on cleaned solutions. On the basis of the results, a possible precipitation route was determined.

Experimental Procedure

Materials

The leach liquor was provided by the General Directorate of Mineral Research & Exploration (MTA), Turkey. The leach liquor was leached from roasted Eskişehir—Beylikova complex ore by 3M hydrochloric acid, 2 h leaching time, and ambient temperature leaching conditions. Oxalic acid (99% w/w) was obtained from Sigma-Aldrich.

Method

Oxalic acid solution was prepared as 1N. A sample of 63 g oxalic acid ($H_2C_2O_4{\cdot}2H_2O$) was dissolved in 1000 mL distilled water and this solution was used for all experiments. Firstly, 100 mL leach liquor was used for experiments. The oxalic acid was added in 10 mL steps, and precipitation was observed at 20 mL oxalic acid addition. First solid/liquid separation was realized at 20 mL oxalic acid addition, and the filter cake was weighed after drying. The precipitate weights were too low for the analysis. The liquor amount was raised to 500 mL for an accurate determination of where precipitation started. For 500 mL leach liquor, oxalic acid volumes were selected as: 20, 40, 60, 150, 200, 250, 300, 400, 500 mL. After the addition of oxalic acid, solutions settled for 15 min. When the last solid/liquid separation was done, the waste solution was settled for 12 h and observed after 12 h.

Characterization

Elemental analysis of the liquor and solutions were realized with Agilent 720 series Radial, Inductively Coupled Plasma—Optical Emission Spectrometry (ICP—OES) and phase analysis of the precipitates were analyzed with Rigaku D/max 2200, X-ray Diffractometer (XRD).

Results and Discussion

Leach Liquor

Initial pH value of the leach liquor was measured as −0.64 and remained below zero for all steps. ICP—OES analysis of HCl leach liquor is given in Table 1.

Table 1 Elemental analysis of the leach liquor

Elements	Ce	La	Nd	Pr	Sm	Th
Conc. (ppm)	16,167	8887	4943	1473	259	410

Table 2 Weight of precipitates

Oxalic Acid Volume (mL)	Weight (g)
20	0.04
40	0.03
60	0.03
100	0.07
150	13.22
200	3.55
250	11.16
300	0.46
400	6.15
500	1.21
12 h settling	0.92

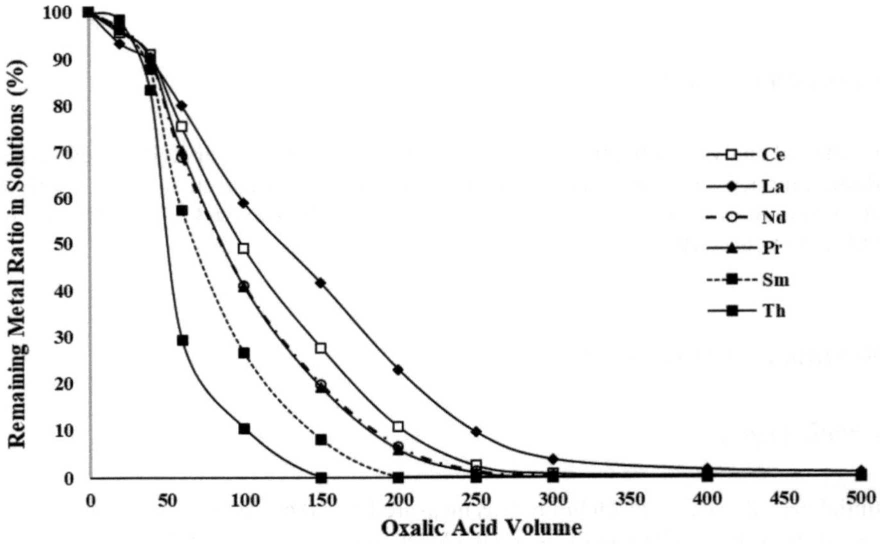

Fig. 1 Remaining metal ratio in solutions versus oxalic acid volume graph

After Precipitation

After precipitation, filter cakes were dried at ambient temperature for 48 h. Dried filter cakes were weighed, and the results are given in Table 2. The two heaviest precipitates were obtained at 150 and 250 mL oxalic acid additions.

The decreasing metal amount vs. oxalic acid volume graphs is given in Fig. 1. The y-axis of Fig. 1 was calculated using Eq. (3).

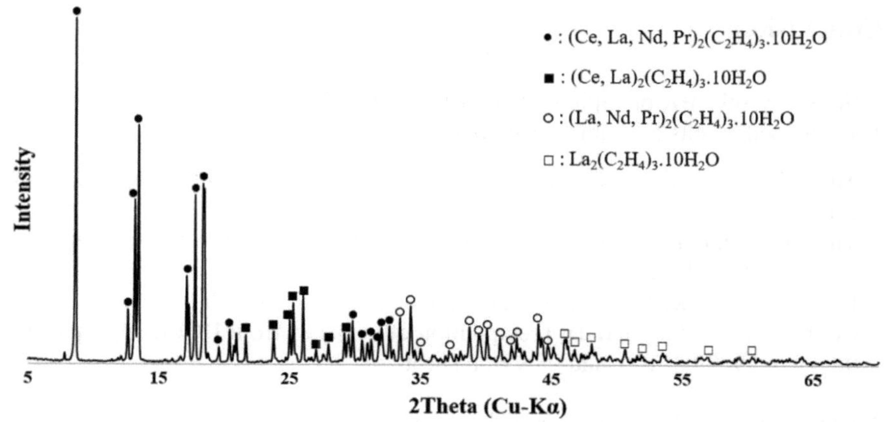

Fig. 2 Phase analysis of the 250 mL oxalic acid addition precipitate

Table 3 Initial and final concentration of the leach liquor

Elements	Ce	La	Nd	Pr	Sm	Th
Final Conc. (ppm)	328	930	49	10	0	4
Efficiency (%)	98	89.5	99	99.3	100	98.9

$$y = \frac{c}{c_i} \times 100 \quad (3)$$

In Eq. (3), c is the present concentration and c_i is the initial concentration of each metal.

As can be seen in Fig. 1, 70% of Th precipitates with the addition of the 60 mL oxalic acid and all Th precipitates with the addition of 150 mL oxalic acid. All Sm was precipitated with 200 mL oxalic acid addition. Nd and Pr were precipitated together. Almost all Ce was precipitated at 250 mL oxalic acid addition, and La was precipitated at 500 mL oxalic acid addition.

Phase analysis of the 250 mL oxalic acid addition is given in Fig. 2. Precipitate with 250 mL oxalic acid addition has four RE-oxalate phases. Precipitates that filtered with under 150 mL oxalic acid addition have not enough amount of solids for XRD analysis.

Elemental analysis of the final solution is given in Table 3. As can be seen, Sm has the highest precipitation efficiency and La has the lowest precipitation efficiency.

Conclusions

In this study, selective precipitation of Th and REEs with oxalic acid was investigated. The obtained results are summarized below:

- The pH value of the leach liquor remained below zero for all investigated oxalic acid additions.
- Th can be separated from REEs with oxalic acid precipitation process with 3/25 oxalic acid/leach liquor volume ratio.
- Precipitation order for the REEs determined as Sm > Nd = Pr > Ce > La.
- Ce, La, Nd, Pr, Sm, and Th precipitation efficiencies from HCl leach liquor are 98, 89.5, 99, 99.3, 100, and 98.9%, respectively.

Acknowledgements The authors would like to acknowledge the General Directorate of Mineral Research & Exploration (MTA), Turkey. This study is supported from The Scientific and Technological Research Council of Turkey (Project No: TÜBİTAK 117M202). This study contains information about one part of this TÜBİTAK project.

References

1. Krishnamurthy N, Gupta CK (2015) Extractive metallurgy of rare earths. CRC Press, Boca Raton
2. Güneş H, Obuz HE, Oğur E, Çapraz F, Alkan M (2018) Rare-earth elements recovery from Nd–Fe–B hard magnets by hydrometallurgical processes. Springer, Cham
3. Obuz HE, Günes H, Kara A, Ugurluer D, Babuccuoglu Y, Alkan M (2018) Leaching kinetics of rare-earth elements from complex ores by acidic solutions. Cham
4. Zhu Z, Pranolo Y, Cheng CY (2015) Separation of uranium and thorium from rare earths for rare earth production—a review. Miner Eng 77:185–196. https://doi.org/10.1016/j.mineng.2015.03.012
5. Zhang J, Zhao B, Schreiner B (2016) Separation hydrometallurgy of rare earth elements. Springer, Berlin
6. Sadri F, Rashchi F, Amini A (2017) Hydrometallurgical digestion and leaching of Iranian monazite concentrate containing rare earth elements Th, Ce, La and Nd. Int J Miner Process 159:7–15
7. Battsengel A, Batnasan A, Narankhuu A, Haga K, Watanabe Y, Shibayama A (2018) Recovery of light and heavy rare earth elements from apatite ore using sulphuric acid leaching, solvent extraction and precipitation. Hydrometallurgy 179:100–109. https://doi.org/10.1016/j.hydromet.2018.05.024
8. Ibrahim TMM, El-Hussaini OM (2007) Production of anhydrite–gypsum and recovery of rare earths as a by-product. Hydrometallurgy 87(1):11–17. https://doi.org/10.1016/j.hydromet.2006.11.017

Part III
Rare Metals III

Experimental Study on the Treatment of Zinc-Containing Rotary Hearth Furnace Dust

Shilei Ren, Xiaoping Liang, Zhongbing Tu, Qian Tang, Xiangguan Yang and Yu Wang

Abstract In this chapter, the treatment process of the rotary hearth furnace (RHF) dust containing Zn, K and Na was studied by experimental method, including the processes of water leaching and addition of NaOH to the filtrate produced by the water leaching process. Water leaching experiment showed that almost all K and Na contents entered the filtrate, while the Zn content in the filtrate was reduced by 7%, and the Cl was reduced by 16%. The experimental of NaOH addition to the filtrate showed that the Zn form of the precipitate was ZnO and $Zn_5(OH)_8Cl_2$ when the pH of the solution was 7.5–13. And when the amount of NaOH is increased from 1 time to 1.5 times of the theoretical consumption, the precipitation rate of Zn is increased from 80.31 to 89.54%, and when the amount of NaOH is increased from 1.5 times to 2 times, the precipitation rate of Zn is 89.54%.

Keywords Rotary hearth furnace dust · Zinc extraction · Leaching
Additive NaOH

Introduction

In the steel production process, hundreds of millions of tons of dust are generated every year. Most of the dust contains harmful metals such as Zn and Pb. Therefore, these dusts must be recycled reasonably. The rotary hearth furnace (RHF) is a device for treating these dusts, and the RHF also discharges dust in its flue while treating metallurgical dust. The RHF dust referred to in this paper refers to the secondary dust of the RHF discharged from the RHF when it is treated with metallurgical dust. If a certain method is used to extract valuable metal elements from these dusts, this can not only reduce the pollution to the environment, but can also realize the secondary utilization of valuable metal resources.

S. Ren · X. Liang (✉) · Z. Tu · Q. Tang · X. Yang · Y. Wang
College of Materials Science and Engineering, Chongqing University, Chongqing 400045, China
e-mail: xpliang@cqu.edu.cn

G. Azimi et al. (eds.), *Rare Metal Technology 2019*, The Minerals, Metals & Materials Series, https://doi.org/10.1007/978-3-030-05740-4_10

The treatment method of the RHF dust varies with the dust composition. Zheng [1] used RHF dust of Ma Steel Plant as the raw material (where the zinc was in the form of ZnO, the content is 47.8%). After leaching with atmospheric pressure sulfuric acid, the iron is removed by goethite-oxidation hydrolysis method, then the copper is replaced by zinc powder, and then the process of extracting zinc by electrowinning is used to treat the zinc extraction from the bottom furnace dust. However, for dust containing K and Na alkali elements, K and Na in the acid immersion liquid cannot be removed after acid leaching, which hinders the extraction and utilization of Zn. Liu [2] used the RHF dust of Ma Steel Plant as the raw material (where the zinc was in the form of ZnO, the content was 45.36%), studied the alkali leaching zinc and electrolysis process to treat the RHF dust, and achieved zinc extraction, but for the dust containing K and Na alkali metal elements, a large amount of K and Na entered the alkali immersion liquid, so that the zinc-containing Zn electrolyte cannot be used in the electrolysis process.

Therefore, for the rotary hearth dust containing K and Na alkali metals, it is necessary to study a new method to separate the K and Na elements while extracting Zn. In this paper, the treatment process of the RHF dust containing Zn, K, and Na was studied by the experimental method.

Materials and Methods

The raw materials used in the experiment are the secondary dust of the RHF produced in the production process of a steel plant, and its XRD diffraction pattern is shown in Fig. 1. The elemental composition of the dust was detected by an inductively coupled plasma-optical emission spectrometer (ICP-OES), and the test results are shown in Table 1. The chemical analysis results of the dust are shown in Table 2. The main

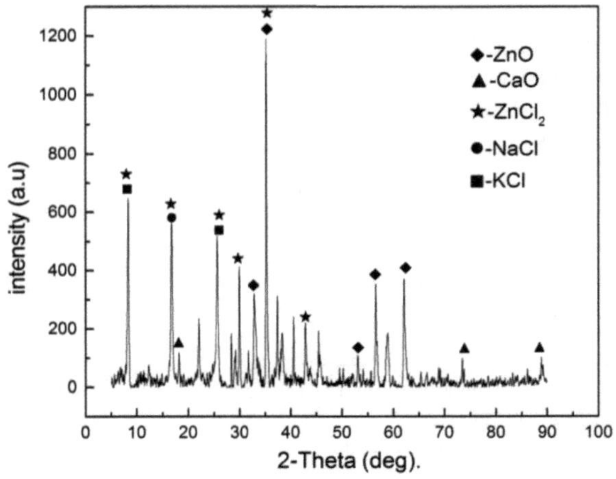

Fig. 1 XRD diffraction pattern of RHF dust

Table 1 Composition and content of RHF dust (wt%)

Ca	Mg	Al	Cu	Fe	Na	K	Zn	Cl	Si
1.00	0.49	0.29	0.15	3.73	7.92	12.77	22.78	47.22	0.68

Table 2 Chemical analysis results of RHF dust

Ingredient	Fe_2O_3	ZnO	CaO	MgO	SiO_2	Cl
Content (wt%)	1.49	6.84	0.39	0.24	0.88	47.22

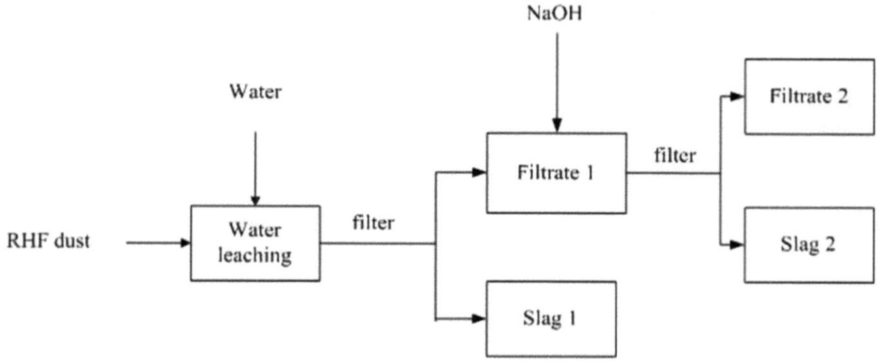

Fig. 2 Experimental flow chart

elements of the dust are Zn, K, Na, and Cl, and it also contains a small amount of impurity elements such as Ca and Mg. Among them, Zn mainly exists in the form of $ZnCl_2$ and a small amount of ZnO, and K and Na exist in the form of KCl and NaCl.

The process of treating RHF dust is shown in Fig. 2.

Water and RHF dust were mixed at a mass ratio of 3:1 and stirred for 60 min and then filtered to obtain filtrate 1 and slag 1. Then, NaOH was added to the filtrate 1 and stirred for 60 min [3–6] and filtered to obtain a filtrate 2 and a slag 2. The composition of the filtrate 1 was detected by ICP-OES. Phase analysis of slag 2 by XRD diffractometer was performed. The formula for calculating the precipitation rate of elements is as follows:

$$\Psi = \frac{M_T \times w_T}{V_T \times k_T \times 10^{-6}} \times 100\% \quad (1)$$

Ψ Element precipitation rate, %;
M_T Quality of filter residue, g;
W_T Element content in filter residue, %;
V_T Initial volume of solution, mL;
k_T Concentration of elements in solution, mg/L;

Phase Equilibrium Analysis of Adding NaOH to Filtrate 1

In this section, the phase equilibrium analysis of the filtrate 1 with NaOH addition was carried out using HYDRA-MEDUSA software. The theoretical NaOH addition amount was determined according to the elemental composition of the filtrate 1 obtained by the water leaching experiment, and then the final ionic component of the filtrate 1 under the three addition conditions was calculated, and the results are shown in Table 3.

Using HYDRA-MEDUSA software, the precipitation of dissolved phase composition of Zn element after the filtrate 1 system reached equilibrium was calculated when the NaOH addition amount was 1 time, 1.5 times and 2 times of the theoretical consumption, respectively, as shown in Figs. 3, 4 and 5.

It can be seen from Figs. 3, 4 and 5 that the difference in the amount of NaOH added has little effect on the phase composition of Zn in the filtrate 1 in which the

Table 3 Phase equilibrium component of the filtrate 1 under three NaOH added conditions

	Fe^{2+}(mM)	Zn^{2+}(mM)	Na^{+}(M)	K^{+}(mM)	Cl^{-}(M)	Mg^{2+}(mM)	Ca^{2+}(mM)
1	58.52	528.52	2.09	723.05	2.81	19.48	34.40
1.5	58.52	538.52	2.74	723.05	2.81	19.48	34.40
2	58.52	1000	3.39	723.05	2.81	19.48	34.40

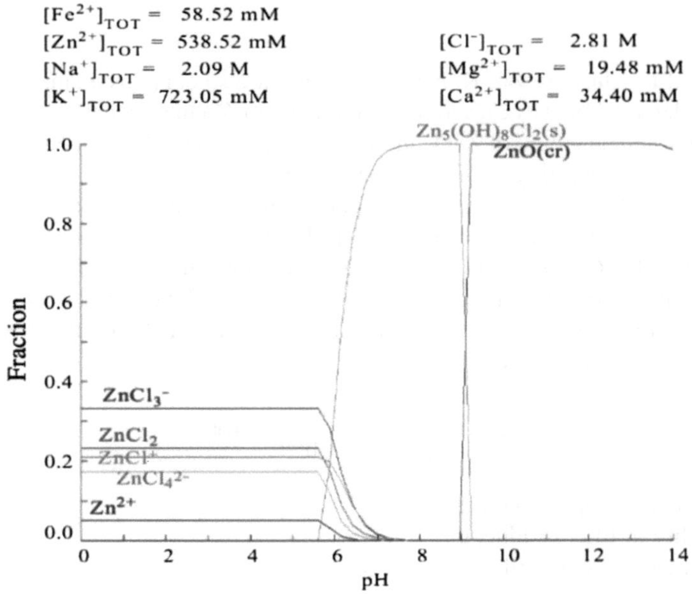

Fig. 3 The phase composition of Zn when the amount of NaOH added is 1 time of the theoretical consumption

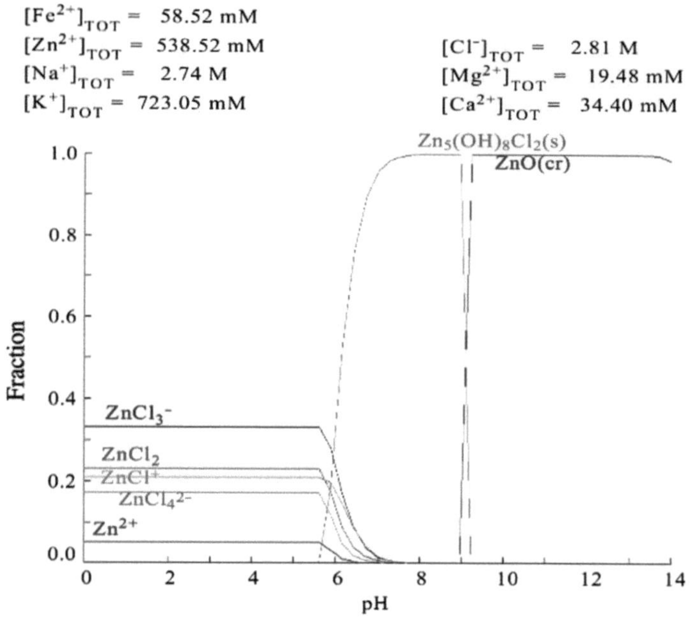

Fig. 4 The phase composition of Zn when the amount of NaOH added is 1.5 times of the theoretical consumption

reaction reaches equilibrium and is only slightly different when the pH of the filtrate 1 is 0–6. Under the condition that the amount of NaOH added is 1 time and 1.5 times of the theoretical consumption, the Zn^{2+} in the filtrate 1 is converted to the $ZnCl^{3-}$ form, and the Zn^{2+} in the filtrate 1 is converted to the $ZnCl^{+}$ form under the condition that the NaOH addition amount is 2 times. In addition, the presence of Zn in the filtrate 1 is basically the same in other pH ranges, that is, the pH of the solution is in the range of 6–9, and the Zn^{2+} in the filtrate 1 is converted into the solid form of $Zn_5(OH)_8Cl_2$. The pH of the solution system is in the range of 9–14, and the Zn^{2+} in the filtrate 1 exists in the form of solid $Zn_5(OH)_8Cl_2$ and ZnO.

Results and Discussion

Water Leaching

The content of elements Zn, K, Na and Cl in RHF dust and filtrate 1 is shown in Fig. 6. The elemental composition of the filtrate 1 is shown in Table 4.

After the process of water leaching, all elements enter filtrate 1 and slag 1. It can be seen from Table 4 and Fig. 6 that almost all K and Na content entered filtrate 1,

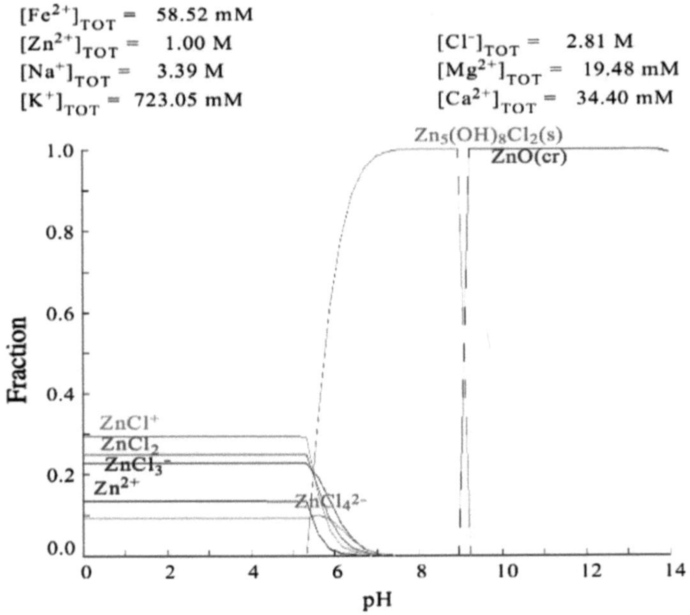

Fig. 5 The phase composition of Zn when the amount of NaOH added is 2 times of the theoretical consumption

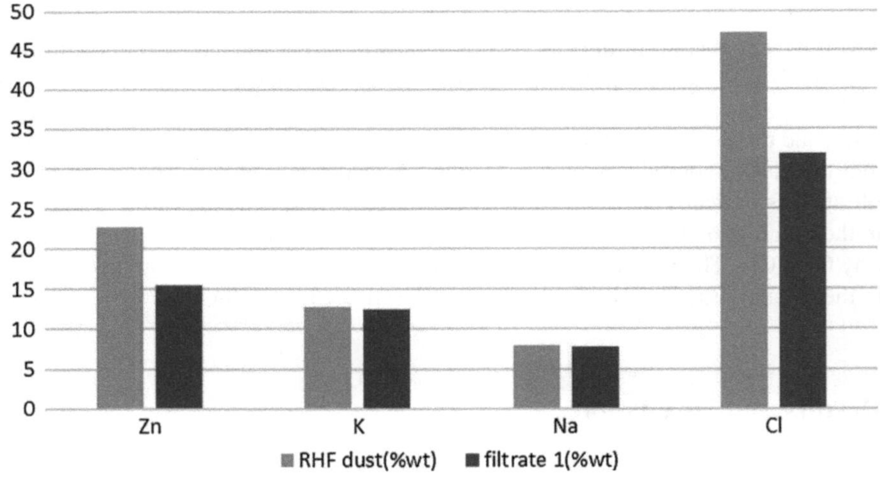

Fig. 6 The content of elements Zn, K, Na and Cl in RHF dust and filtrate 1

Table 4 Elemental composition and content of the filtrate 1

Ingredient	K	Na	Zn	Ca	Mg	Fe	Cl
Content (mol/L)	0.723	0.784	0.538	0.034	0.019	0.058	2.807

while the Zn content in the filtrate 1 was reduced by about 7%, and the Cl element content was reduced by about 16%. The reduction in the Zn content may be due to a small portion of ZnO in the raw material that is insoluble in water. The decrease in Cl may be due to a portion of the solution remaining in the slag 1 when filtered.

Addition of NaOH to Filtrate 1

The effect of different NaOH additions on the precipitation rate of Zn in the filtrate 1 was studied at 60 min precipitation residence time. The experimental results on the precipitation rate of Zn are shown in Table 5 and Fig. 7. The XRD pattern of the resulting precipitate is shown in Figs. 8, 9 and 10.

Table 5 Experimental results under different NaOH addition

NaOH addition amount (relative to multiples of theoretical consumption)	Zn precipitation rate (%)	pH
1	80.31	7.5
1.5	89.54	11
2	90.29	13

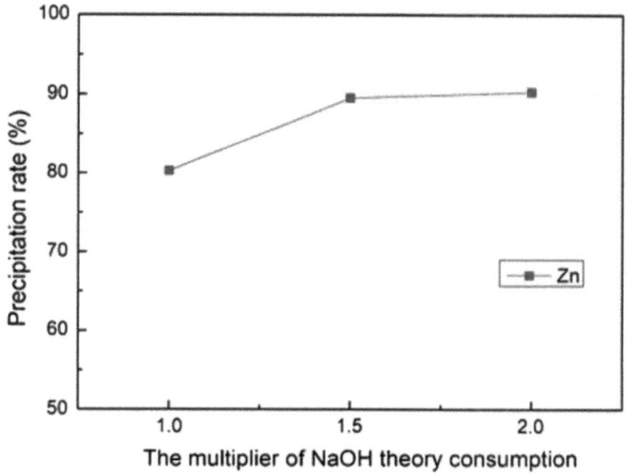

Fig. 7 The effect of different NaOH addition on the precipitation rate of Zn

From Table 5 and Fig. 7, as the amount of NaOH added increases, the precipitation rate of Zn gradually increases. When the amount of NaOH added was increased from 1 time to 1.5 times of the theoretical consumption, the precipitation rate of Zn in the filtrate 1 extract increased from 80.31 to 90.29%, increased by 9.98%; at the same time, the pH increased from 7.5 to 11. When the amount of NaOH added is increased from 1.5 times to 2 times, the increase in Zn precipitation rate becomes slow, the growth rate is only 0.75%, and the pH is increased from 11 to 13.

The experimental results of adding NaOH to the filtrate 1 showed that the pH of the solution was 7.5, 11 and 13, respectively, when the NaOH addition amount was 1 time, 1.5 times and 2 times of the theoretical consumption. It can be seen from Figs. 8, 9 and 10 that $Zn_5(OH)_8Cl_2$ and ZnO are present in the slag obtained under different NaOH addition amounts, indicating that when the pH of the solution is in the range of 7.5–13, the reaction of NaOH with the filtrate 1 simultaneously generates ZnO and $Zn_5(OH)_8Cl_2$. According to the results of solution phase equilibrium analysis, when the pH of the solution system is in the range of 6–9, slag contains $Zn_5(OH)_8Cl_2$.

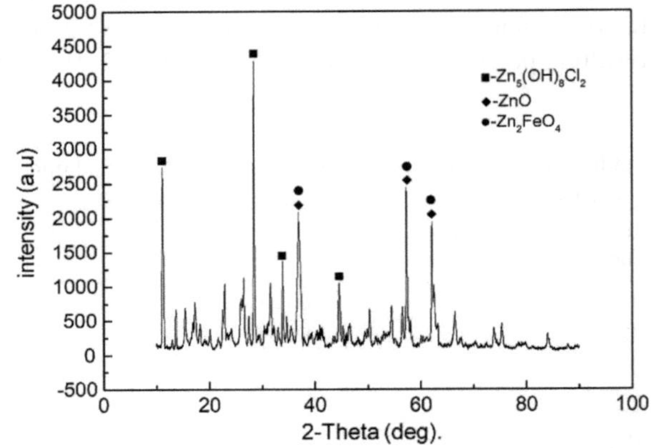

Fig. 8 The XRD diffraction pattern of the slag 2 when the amount of NaOH was 1 time of the theoretical consumption

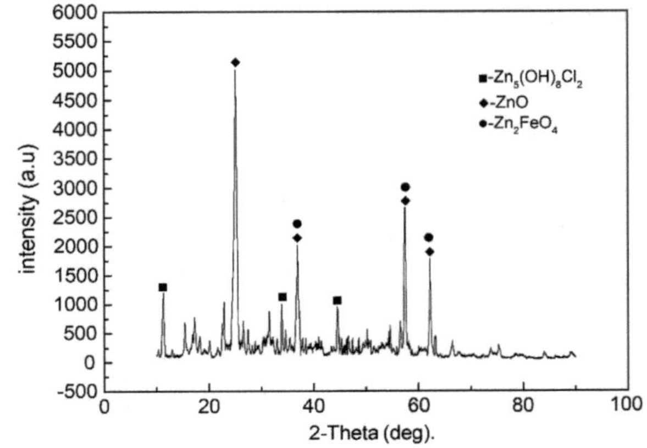

Fig. 9 The XRD diffraction pattern of the slag 2 when the amount of NaOH was 1.5 times of the theoretical consumption

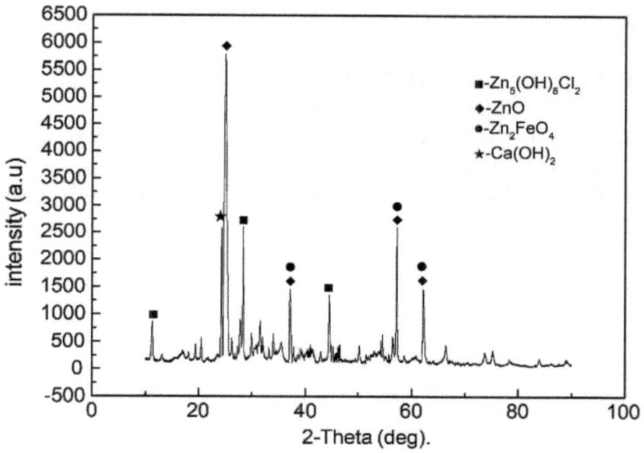

Fig. 10 The XRD diffraction pattern of the slag 2 when the amount of NaOH was 2 times of the theoretical consumption

When the pH of the solution system is in the range of 9–14, the obtained slag contains ZnO and $Zn_5(OH)_8Cl_2$. Therefore, the experimental results are basically consistent with the theoretical analysis of phase equilibrium.

Conclusions

The water leaching experiment showed that almost all K and Na content of the feed entered filtrate 1, while the Zn content in the filtrate 1 was reduced by about 7%, and the Cl content was reduced by about 16%. The phase equilibrium analysis on adding NaOH to the filtrate showed that when the pH of the solution was 6–9, the form of Zn in the precipitate was $Zn_5(OH)_8Cl_2$, and when the pH was 9–14, the form of Zn in the precipitate was ZnO and $Zn_5(OH)_8Cl_2$. The experimental results of adding NaOH to the filtrate 1 showed that the Zn form of the precipitate (slag 2) was ZnO and $Zn_5(OH)_8Cl_2$ when the pH of the solution was 7.5–13, which was consistent with the phase equilibrium analysis. In addition, when the NaOH addition is increased from 1 time to 1.5 times of the theoretical consumption, the Zn precipitation rate is increased from 80.31 to 89.54%, and when the NaOH addition is increased from 1.5 times to 2 times, the Zn precipitation rate is increased from 89.54 to 90.29%.

References

1. Zheng W (2012) Technology research of acid treatment of rotary hearth furnace dust containing rich zinc and lead content. Anhui University of Technology, China
2. Liu H (2012) Technology research of rotary hearth furnace rich lead-zinc ash by alkaline processing. Anhui University of Technology, China

3. Wang D, Jiao X, Chen D (2003) Brief Introduction on properties, synthesis and applications of zinc sulfide. ShanDong Chem Industry 32:12–15
4. Lenz DM, Martins FB (2007) Lead and zinc selective precipitation from leach electric arc furnace dust solutions. Revista Materia 12(3):503–509
5. Peng H, Yao Y, Deng M (2015) Study on the synthesis of basic zinc chloride from zinc chloride and sodium hydroxide. Feed Wide Angle 12:23–26
6. Chen W, Xue Y (2011) Preparation of nano-basic sodium carbonate with different particle sizes. Liaoning Chem Industry 40(1):7–8

Recovery of Manganese by Roasting-Ammonia Leaching from Low-Grade Manganese Carbonate Ores

Zhongbing Tu, Xiaoping Liang, Xiangguan Yang, Shilei Ren, Chengbo Wu and Yu Wang

Abstract The utilization of low-grade manganese ores has drawn much attention owing to the rapid depletion of high-grade manganese ore resources. In this study, low-grade manganese carbonate ores were treated by roasting-ammonia leaching method. The effects of roasting temperature, roasting time, ammonia concentration, ammonium carbonate concentration and liquid-to-solid ratio on Mn leaching were discussed. The leaching efficiency of 85.6 for Mn was obtained under the following conditions: 650 of roasting temperature, 1 h roasting time, 14 mol/L of ammonia concentration, 2 mol/L of ammonium carbonate concentration, 80 min of leaching time and 5:1 of liquid/solid ratio.

Keywords Low-grade ores · Manganese carbonate · Ammonia · Leaching

Introduction

High-grade manganese ores (> 40%) are typically processed into suitable metallic alloy forms by pyrometallurgical processes [1]. The low-grade manganese ores (< 40%) , which do not find any use, are usually dumped at the mine site [2]. With the increasing demand for manganese products, low-grade manganese ores become important sources of manganese. Hydrometallurgical treatment of low-grade manganese ores attracts the attention of researchers in recent years.

Many studies have been done to investigate recovery of manganese from low-grade manganese ores by hydrometallurgical processes. When manganese is present in its bivalent soluble form Mn (II), manganese salts are generally obtained directly by acid leaching; when manganese is present in manganese dioxide, which is insoluble in dilute acid media, it is necessary to convert manganese dioxide into Mn(II) compounds by reduction roasting followed by acid leaching [3, 4] or direct reductive

Z. Tu · X. Liang (✉) · X. Yang · S. Ren · C. Wu · Y. Wang
College of Materials Science and Engineering, Chongqing University, 400045 Chongqing, China
e-mail: xpliang@cqu.edu.cn

G. Azimi et al. (eds.), *Rare Metal Technology 2019*, The Minerals, Metals & Materials Series, https://doi.org/10.1007/978-3-030-05740-4_11

acid leaching using different kinds of acids and reducing agents [5–8]. For instance, gaseous sulfur dioxide, carbon, cornstalk and sulfur have been investigated as reductants for pyrometallurgical pre-treatment followed by acid leaching [9–12]. Moreover, manganese can also be extracted directly by reductive acid leaching, employing diverse reducing agents, such as pyrite, hydrogen peroxide, sulfite, iron metal, iron (II) sulfate, oxalic acid, citric acid, glucose. [13–16]. Satisfactory manganese leaching efficiency is obtained in these acid leaching methods. But most of these methods have not found commercial application for various reasons, including high acid consumption, equipment erosion and the inability to recycle leach solutions. Therefore, a low-cost and environmental process for extracting manganese is being sought.

Attributing to its low cost and ease of regeneration, ammonia has been investigated as an effective lixiviant in leaching some manganiferous slags [17, 18]. The flow sheet of roasting-ammonia leaching process is schematically illustrated in Fig. 1. $NH_3{\cdot}H_2O$ and $(NH_4)_2CO_3$ are used as leaching agent in leaching process, and after distillation, NH_3 and CO_2 are released from solution and then absorbed by water to form $NH_3{\cdot}H_2O$ and $(NH_4)_2CO_3$ again. Most leaching agent is recycled and barely waste liquor is generated in roasting-ammonia leaching process which make this process environmentally friendly.

In order to utilize low-grade manganese ores efficiently and environmentally, this study investigated the roasting-ammonia leaching method to extract manganese from low-grade manganese ores. The effects of important parameters including roasting temperature, roasting time, ammonia concentration, ammonium carbonate concentration and liquid-to-solid ratio on the manganese leaching efficiency were examined.

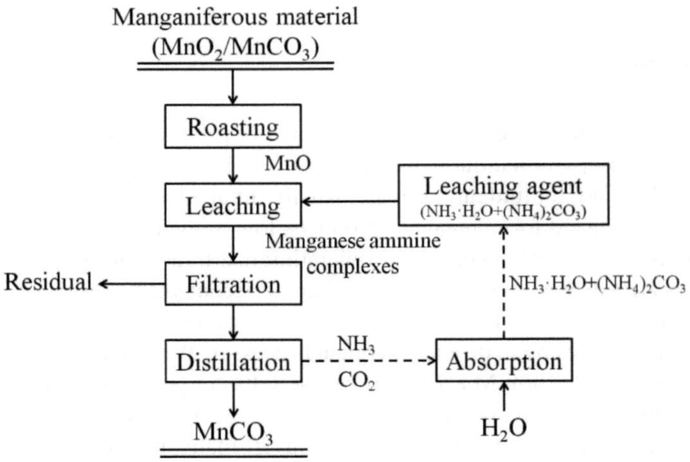

Fig. 1 Flow sheet of ammonia leaching process [19]

Material and Method

Material

The low-grade manganese ore used in this study was obtained from Wan Yuan, Sichuan Province, China. The chemical composition and X-ray diffraction pattern are given in Table 1 and Fig. 2, respectively. The material is a typical low-grade manganese carbonate ore, and the mass fraction of manganese is only 17.83%. The XRD pattern shows the ore is mainly composed of rhodochrosite, dolomite, ankerite and a small quantity of quartz.

Experimental Procedure

At first, the low-grade manganese carbonate ores were roasted in a tube furnace under a flow of nitrogen. The manganese monoxide was obtained by roasting manganese carbonate according to the following reaction:

$$MnCO_3 = MnO + CO_2 \tag{1}$$

After roasting was completed, the roasted product was placed into a three-neck spherical glass reactor equipped with a mechanical stirrer, a temperature control unit and a cooler to avoid evaporation loss of solution and ammonia. Specific concentra-

Table 1 Chemical compositions of low-grade rhodochrosite (mass fraction, %)

Mn	Ca	Si	Mg	Fe	S	Al	P
17.83	12.60	3.05	4.25	1.24	0.51	1.30	0.40

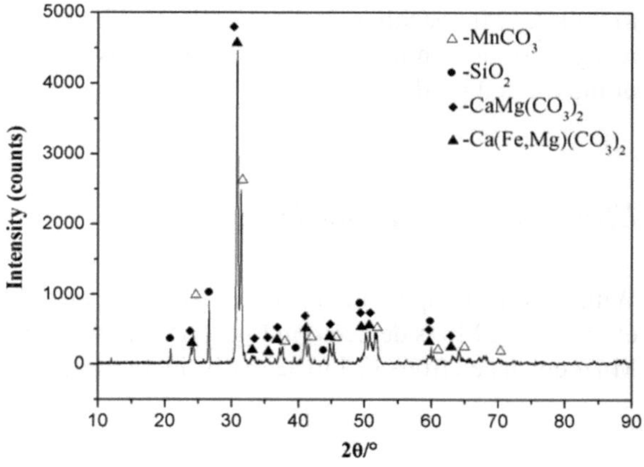

Fig. 2 XRD pattern of low-grade manganese ore

tion of ammonia and ammonium carbonate was also placed into the glass reactor. The main reaction in leaching process can be expressed as:

$$2NH_3 + CO_2 = NH_2COO^- + NH_4^+ \quad (2)$$

$$MnO + xNH_3 + NH_2COO^- = [MnO(NH_3)_xNH_2COO]^- \quad (3)$$

The solution was stirred continuously. After the sample was filtrated, manganese in residue was detected by perchloric acid ammonium ferrous sulfate titration [20]. The leaching recovery of manganese can be calculated by the following formula:

$$\phi = 1 - \frac{\omega \times m_1}{17.83 \times m_2} \times 100\%$$

Φ Manganese recovery, %;
ω Mass fraction of residue, %;
m_1 Mass of residue, g;
m_2 Mass of raw material, g.

Results and Discussion

Effect of the Roasting Temperature

The effect of roasting temperature on the leaching efficiency of Mn is presented in Fig. 3. With raising the roasting temperature from 500 to 650 °C, the manganese recovery has a sharp increase from 4.9 to 83.5%. As reported, MnO instead of $MnCO_3$ can be dissolved in ammonia–ammonium carbonate solution. In this stage, higher Mn leaching efficiency for higher roasting temperature is due to more $MnCO_3$ decomposed into MnO. However, when the roasting temperature continues to increase, leaching efficiency of Mn tends to decrease. When the roasting temperature increases to 900 °C, Mn leaching efficiency falls to 72.3%. That is the result of over-heating manganese ores in the furnace, which affects the reactivity and phase composition of the roasted product. Therefore, over-heating of ore sample should be prevented.

Effect of the Roasting Time

With roasting temperature fixed as 650 °C, the effect of roasting time on the leaching efficiency of Mn is depicted in Fig. 4. As shown in Fig. 4, the leaching efficiency of Mn is enhanced from 32.1 to 82.6% by increasing the roasting time from 20 to 80 min.

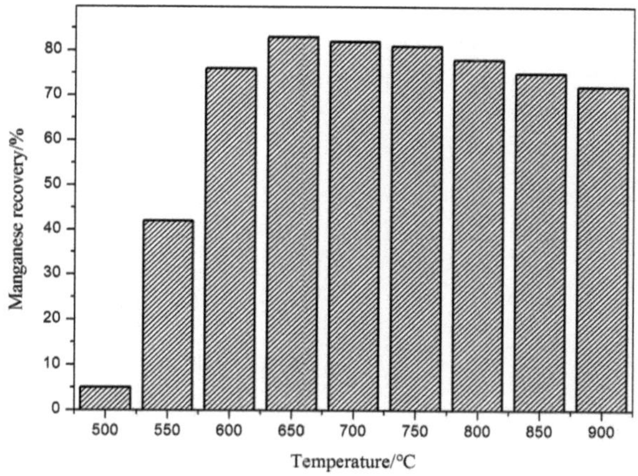

Fig. 3 Effect of roasting temperature on manganese recovery

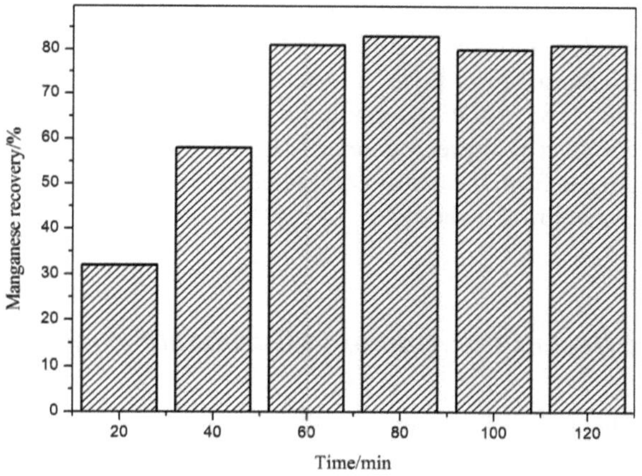

Fig. 4 Effect of roasting temperature on manganese recovery

However, when the roasting time rises from 80 to 120 min, Mn leaching efficiency decreases slightly from 82.6 to 81.2%, which demonstrates that the decomposition reaction is close to equilibration at 80 min. Therefore, taking the roasting cost into consideration, the most favorable roasting time should be 80 min.

Effect of the Ammonia Concentration

The effect of ammonia concentration on the leaching efficiency of Mn is given in Fig. 5. The leaching efficiency curves in Fig. 5 show that Mn leaching efficiency increases significantly from 15.5 to 84.1% with increasing the ammonia concentration from 8 to 14 mol/L. When ammonia concentration continues to increase from 14 to 16 mol/L, Mn leaching recovery was almost unchanged. This is because of the

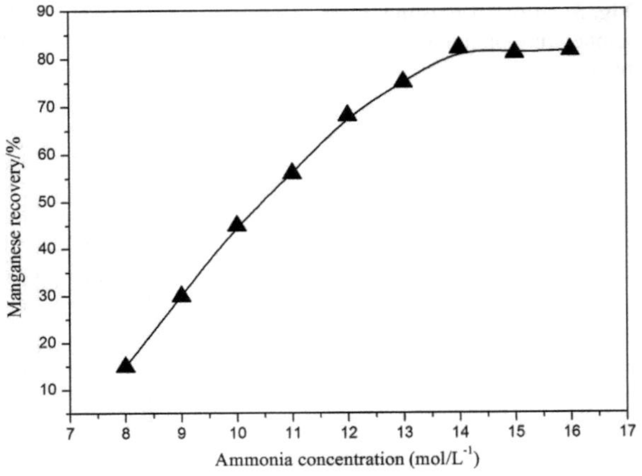

Fig. 5 Effect of ammonia concentration on manganese recovery

weak combination ability of Mn with NH_3 to form manganese amine complexes; leaching Mn requires an excess of ammonia to increase the stability of the complex [18, 21]. When ammonia concentration is not high enough, manganese amine complex will precipitate as $MnCO_3$ into residue. Therefore, in order to leach manganese from the ore as completely as possible and reduce cost, the optimal ammonia concentration should be 14 mol/L.

Effect of the Ammonia Carbonate Concentration

The effect of ammonia carbonate concentration on Mn extraction was also studied, and the results are given in Fig. 6. Figure 6 indicates that the extraction efficiencies of Mn rise with the increase of the ammonia carbonate concentration. With increasing the ammonia carbonate concentration from 0 to 2 mol/L, the Mn leaching efficiency augments from 7.1 to 81.3%. When ammonia carbonate concentration continues to increase from 2 to 3 mol/L, Mn leaching recovery slightly increases from 81.3 to 83.6%. It is reported that addition of ammonia carbonate in ammonia solution is to adjust pH of the solution so that manganese amine complex can be stable; the other function is offering monovalent anion to form manganese amine complex [21]. The increase of the ammonia carbonate concentration over 2 mol/L has no little influence on the Mn leaching efficiency; the suitable ammonia carbonate concentration should be 2 mol/L.

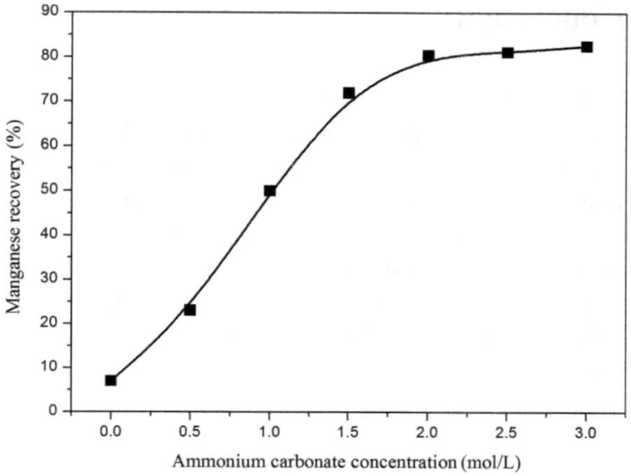

Fig. 6 Effect of ammonia concentration on manganese recovery

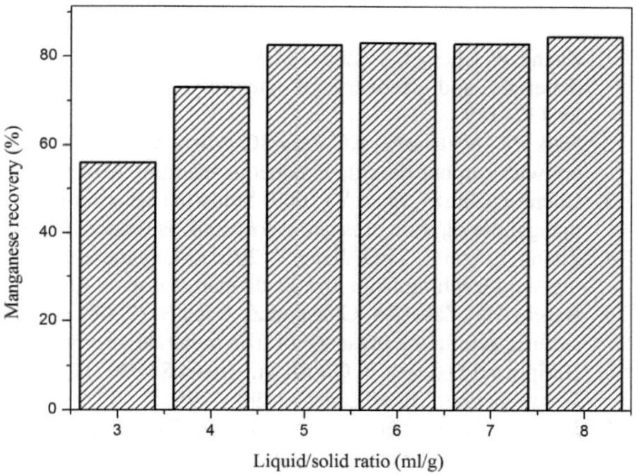

Fig. 7 Effect of liquid-to-solid ratio on manganese recovery

Effect of Liquid-to-Solid Ratio

The effect of liquid/solid ratio on the manganese recovery was investigated by varying the ratio from 3:1 to 8:1. The results are shown in Fig. 7. With increasing the liquid/solid ratio from 3:1 to 5:1, the Mn recovery rises from 56.5 to 80.6%. At the liquid/solid ratio of 3:1, Mn extraction is too low because the ammonia and ammonium carbonate amount is insufficient to dissolve most of manganese. Under the condition of a fixed ammonia and ammonium carbonate concentration, the bigger liquid/solid ratio means more free ammonia and ammonium ions in the solution, which accelerates leaching reaction. When the liquid/solid reaches to 5:1, most manganese has been leached. After that, the increase of the ratio has no significant effect on the Mn leaching efficiency. Hence, the liquid/solid ratio of 5:1 is recommended in this investigation.

Conclusions

A novel process for recovering manganese from low-grade manganese carbonate ores by roasting-ammonia leaching method was investigated in this paper. Based on single-factor analysis method, the optimal condition for roasting-ammonia leaching manganese from low-grade manganese carbonate ores was determined as roasting temperature of 650 °C for 80 min, 14 mol/L of ammonia concentration, 2 mol/L of ammonium carbonate concentration and 5:1 of liquid/solid ratio. Under the optimal condition, the leaching recovery of Mn can reach 85.6%. Results of the investigations indicated that relatively good levels of manganese extraction could be attained by roasting-ammonia leaching method.

References

1. Zhang W, Cheng CY (2007) Manganese metallurgy review. Part I: leaching of ores/secondary materials and recovery of electrolytic/chemical manganese dioxide. Hydrometallurgy 89:137–159
2. Naik PK, Nathsarma KC, Das SC, Misra VN (2003) Leaching of low grade Joda manganese ore with sulphur dioxide in aqueous medium. Mineral Process Extr Metall 112:131–134
3. Cheng Z, Zhu G, Zhao Y (2009) Study in reduction-roast leaching manganese from low-grade manganese dioxide ores using cornstalk as reductant. Hydrometallurgy 96:176–179
4. Zhang Y, You Z, Li G, Jiang T (2013) Manganese extraction by sulfur-based reduction roasting–acid leaching from low-grade manganese oxide ores. Hydrometallurgy 133:126–132
5. Bafghi MS, Zakeri A, Ghasemi Z, Adeli M (2008) Reductive dissolution of manganese ore in sulfuric acid in the presence of iron metal. Hydrometallurgy 90:207–212
6. Hariprasad D, Dash B, Ghosh M, Anand S (2007) Leaching of manganese ores using sawdust as a reductant. Miner Eng 20:1293–1295
7. Jiang T, Yang Y, Huang Z, Zhang B, Qiu G (2004) Leaching kinetics of pyrolusite from manganese–silver ores in the presence of hydrogen peroxide. Hydrometallurgy 72:129–138
8. Nayl A, Ismail I, Aly H (2011) Recovery of pure $MnSO_4{\cdot}H_2O$ by reductive leaching of manganese from pyrolusite ore by sulfuric acid and hydrogen peroxide. Int J Miner Process 100:116–123
9. Welham NJ (2002) Activation of the carbothermic reduction of manganese ore. Int J Miner Process 67:187–198
10. Zhang Y, You Z, Li G, Jiang T (2013) Manganese extraction by sulfur-based reduction roasting–acid leaching from low-grade manganese oxide ores. Hydrometallurgy 133:126–132
11. Freitas LR (1983) Sulphation of Carajás manganese ore with gaseous SO_2. Trans Institution Mining Metall vol Tech notes:C130–C131
12. Cheng Z, Zhu G, Zhao Y (2009) Study in reduction-roast leaching manganese from low-grade manganese dioxide ores using cornstalk as reductant. Hydrometallurgy 96:176–179
13. Nayak BB, Mishra KG, Paramguru RK (1999) Kinetics and mechanism of MnO_2 dissolution in H_2SO_4 in the presence of pyrite. J Appl Electrochem 29:191–200
14. Furlani G, Pagnanelli F, Toro L (2006) Reductive acid leaching of manganese dioxide with glucose: identification of oxidation derivatives of glucose. Hydrometallurgy 81:234–240
15. Bafghi MS, Zakeri A, Ghasemi Z, Adeli M (2008) Reductive dissolution of manganese ore in sulfuric acid in the presence of iron metal. Hydrometallurgy 90:207–212
16. Su H, Wen Y, Wang F, Sun Y, Tong Z (2008) Reductive leaching of manganese from low-grade manganese ore in H_2 SO_4 using cane molasses as reductant. Hydrometallurgy 93:136–139

17. Heindl R, Ruppert J, Skow M, Conley J (1955) Manganese from steel-plant slags by a lime-clinkering and carbonate-leaching process: part I, laboratory development (in two parts). BuMines Rep Inv 5124:98
18. Meng X, Han KN (1996) The principles and applications of ammonia leaching of metals—a review. Mineral Process Extr Metull Rev 16:23–61
19. Mcintosh SN, Baglin EG (1992) Recovery of manganese from steel plant slag by carbamate leaching. US Department of the Interior, Bureau of Mines
20. Liu Y, Lin Q, Li L, Fu J, Zhu Z, Wang C et al (2014) Study on hydrometallurgical process and kinetics of manganese extraction from low-grade manganese carbonate ores. Int J Mining Sci Technol 24:567–571
21. Devuyst E (1970) Reductive dissolution of goethite and pyrolusite in alkaline solution. University of British Columbia, British Columbia, Canada

General Rules for Deep Purification of Low-Grade Molybdenite Concentrates

Junjie Yu, Hu Sun, Jun Luo, Guanghui Li and Tao Jiang

Abstract Low-grade molybdenum concentrate is a typical resource to extract molybdenum, but it has long been used inefficiently due to its high contents of impurities such as Ca, Fe, Pb, and Cu. In this study, the deep purification of low-grade molybdenite concentrates have been investigated via both theory and experiments. E-pH calculation shows that acid system is beneficial to the leaching of impurities. When the pH is around zero, the oxidation potential for gangue minerals to dissolve follows the order: E (Cu_2S) > E (PbS) > E (FeS). Subsequent purification tests indicate that oxidants such as $FeCl_3$ and $CuCl_2$ could intensify the removal of impurities but also contribute to the increase of Re loss. Thus, precise control of operating variables plays an important role in balancing the removal of impurities with Re reservation.

Keywords Low-grade molybdenite concentrate · Impurities · E-pH · Purification

Introduction

Molybdenum (Mo) is an essential element that is widely used in the steel industry, chemical industry, non-ferrous metallurgy, and agriculture [1]. By 2015, the global molybdenum production had reached 235,000 metric tons, which is sixteen times the level in 1950 [2]. As the demand for molybdenum is increasing rapidly, the high-grade molybdenite concentrate used to extract molybdenum is diminishing, and it is necessary to make full use of low-grade molybdenite concentrate. However, until now, low-grade molybdenite concentrate has not been effectively used, due to its high contents of impurities such as Ca, Fe, Pb, and Cu. Usually, these impurities mainly exist in form of minerals such as calcite, pyrite chalcocite, chalcopyrite, and galena. During the oxidative roasting process, mainly used to recover molybdenum,

J. Yu · H. Sun · J. Luo · G. Li (✉) · T. Jiang
School of Minerals Processing and Bioengineering, Central South University, Changsha 410083, Hunan, China
e-mail: liguangh@csu.edu.cn

G. Azimi et al. (eds.), *Rare Metal Technology 2019*, The Minerals, Metals & Materials Series, https://doi.org/10.1007/978-3-030-05740-4_12

Table 1 Main chemical composition of molybdenite concentrate sample (wt%)

Mo	Re	Fe	SiO_2	Al_2O_3	CaO	Pb	Cu	S
43.55	321 g/t	1.39	4.02	0.18	3.27	4.52	0.026	32.46

the above gangues are transformed into oxides and react with molybdenum oxide to form molybdate, which may reduce the recovery rate of molybdenum [3, 4]. In addition, impurities such as lead can volatilize in later roasting process, which would pose a great threat to our environment. In order to make full use of the low-grade molybdenite concentrates, much research about impurities removing has been done. Typically, hydrochloric acid was adopted to purify lead-rich molybdenite concentrate, and our former research revealed that 93.6% Pb could be removed with HCl [5]. Besides, brine containing ferric chloride was also reported to be effective to remove Cu and Pb from molybdenite concentrate [6].

However, up to now, systematic studies on specific influence of various purification agents on the removal of each impurity component and the loss of molybdenum and rhenium are still lacking. General rules for purifying different types of low-grade molybdenite concentrates are urgently to be established. Herein, E-pH diagrams of ReS_2, MoS_2, Cu_2S, PbS, and FeS under the hydrochloric acid system were plotted first to obtain the appropriate conditions for deep purification of low-grade molybdenite concentrates. Then, hydrochloric acid and chlorides including $CaCl_2$, $CuCl_2$, and $FeCl_3$ were adopted to explore the removal of impurities from a low-grade and rhenium-bearing molybdenite concentrate. Finally, general rules for deep purification of low-grade molybdenite concentrates were summarized, and the relationship between the removal of impurities and the loss of molybdenum and rhenium was also revealed.

Experimental

Materials

Molybdenite concentrate used in the experiments was provided by a dressing plant in Shanxi Province. Chemical compositions and X-ray diffraction (XRD) pattern of the raw material were shown in Table 1 and Fig. 1, respectively. The molybdenite concentrate comprises 43.55 wt% Mo, belonging to low-grade molybdenum ore. The rhenium in concentrate reaches 321 g/t, which demonstrate the great value for recovering Re. The X-ray pattern shows that molybdenite is the major phase; minor amounts of quartz (SiO_2), galena (PbS) and calcite ($CaCO_3$) are also present. Additionally, hydrochloric acid (36 wt%), $CuCl_2 \cdot 2H_2O$, $FeCl_3 \cdot 6H_2O$, and $CaCl_2 \cdot 2H_2O$ with analytical grade were used as the purification reagents. Double distilled deionized water was used for all tests in the laboratory.

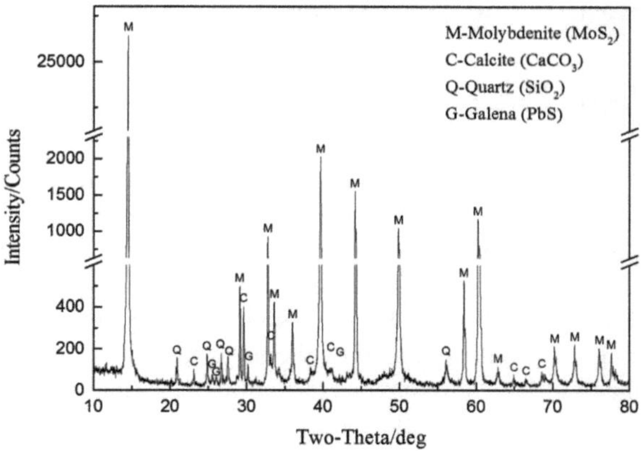

Fig. 1 XRD pattern of the molybdenite concentrate sample

Methods

At first, the molybdenite concentrate was ground to 100 wt% undersize 0.074 mm. In each leaching test, 5 g of the sample was treated in a beaker containing a predetermined quantity of mixed solution [the quantity of mixed solution was determined according to the liquid–solid ratio (mL/g)]. The beaker was located in a water bath that maintains the leaching temperature with a solution stirred continuously. After leaching for a desired duration, the solution was filtered and the residue obtained was washed with hot distilled water and dried for subsequent analysis. The content of rhenium was measured by ICP, while other species were determined by chemical analysis. In addition, the construction of the E–pH diagrams was implemented using thermochemical software HSC chemistry 6.0 (Outokumpu, Helsinki, Finland), and thermodynamic data were all originated from the software itself.

The acidic leaching ratio of Ca, Fe, Pb, Cu, Mo or Re was calculated by the equation as follows:

$$\beta = 1 - \frac{m_1 \times \varepsilon_1}{m_0 \times \varepsilon_0} \times 100\%$$

where β is the leaching ratio, %; m_0 is the weight of molybdenite concentrate, g; ε_0 is the weight fraction of Ca, Fe, Pb, Cu, Mo or Re of molybdenite concentrate, wt%; m_1 is the weight of leached residue, g; and ε_1 is the Pb, Mo or Re weight fraction of leached residue, wt%.

Results and Discussion

E-pH Diagrams of Me_2S–MeCl–H_2O System

The E-pH diagrams of impurities were calculated under the conditions of leaching system of hydrochloric acid, temperature at 95 °C, and molality of chlorine is 1 mol/kg H_2O [5]. As can be seen from Fig. 2a, b and c, the components of PbS, Cu_2S and FeS cannot transfer into solutions, when the pH value is above 10, 3 and 4, respectively. Besides, it is obvious that acid system is propitious to remove impurities. To confirm the optimal conditions for simultaneously leaching Pb, Cu and Fe, the first three diagrams were combined, as shown in Fig. 2d. The shadow area representing the co-dissolution zone clearly shows that, when the pH ranges from −2 to 2, increasing the oxidation potential, components of Cu_2S, PbS and FeS can transfer into solutions simultaneously. Besides, when the pH is around zero, the oxidation potential for impurities to dissolve follows the order: E (Cu_2S) > E (PbS) > E (FeS).

On the other hand, the results are shown in Fig. 3a and b indicate that, as the pH increasing, the oxidation potential needed to dissolve MoS_2, ReS_2 decreases, that is, alkaline system would lead to much greater loss of molybdenum and rhenium.

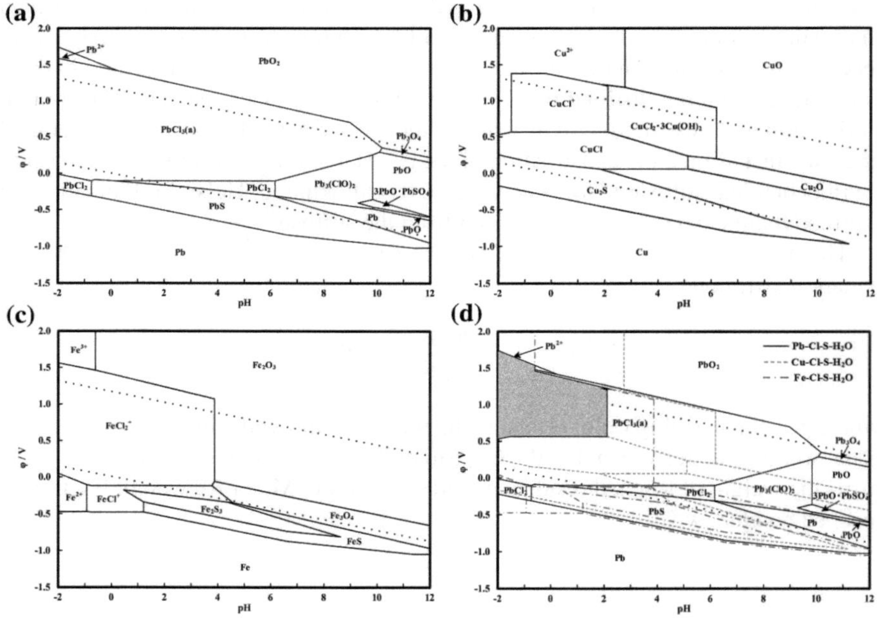

Fig. 2 Potential-pH diagrams of **a** Pb–Cl–S–H_2O system, **b** Cu–Cl–S–H_2O system, **c** Fe–Cl–S–H_2O system, **d** Pb/Cu/Fe–Cl–S–H_2O system at 95 °C

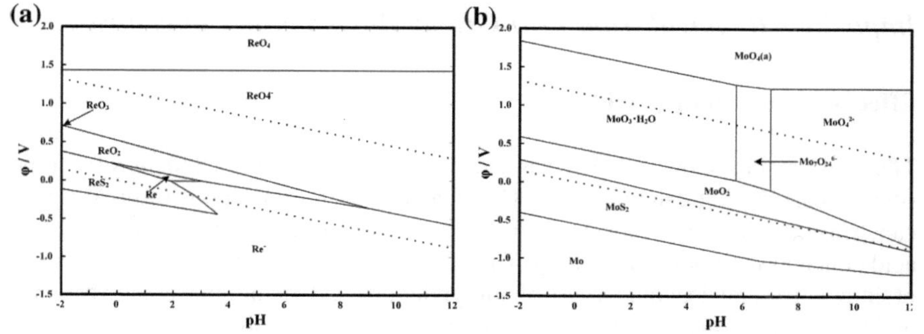

Fig. 3 Potential-pH diagrams of **a** Re–Cl–S–H_2O system, **b** Mo–Cl–S–H_2O system at 95 °C

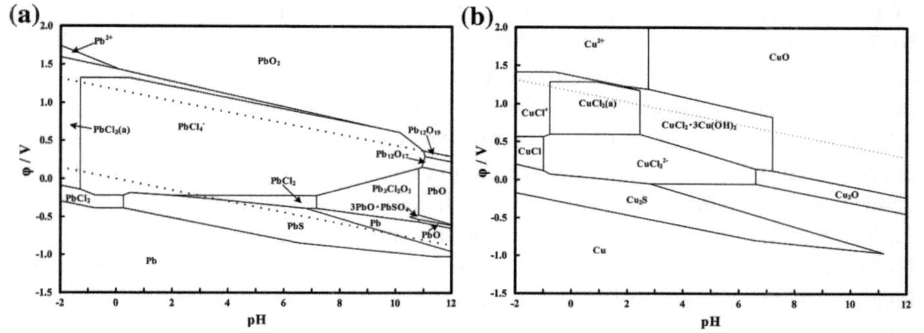

Fig. 4 Potential-pH diagrams of **a** Pb–Cl–S–H_2O system, **b** Cu–Cl–S–H_2O system at 95 °C (Molality: $[Cl^-] = 10$ mol/kg H_2O)

Therefore, acid system (pH < 2) is beneficial to the removing of impurities and the retaining of molybdenum and rhenium.

In addition, it can be seen from Fig. 2a, b and c, that iron–chlorine complexes ($FeCl^+$, $FeCl_2{}^+$) are easy to dissolve in medium or acid water solution, while different complexes of copper–chlorine and lead–chlorine (CuCl, $CuCl_2$, $PbCl_2$, $PbCl_3$) show a great difference in solubility. When PbS and Cu_2S reacting with HCl, the $PbCl_2$(s) and CuCl(s) are forming at first, which can react with Cl^-continually to dissolve later. Thus, the concentration of chlorine has a certain effect on the leaching of PbS and Cu_2S. Given this, E-pH diagrams of PbS and Cu_2S in high concentration (10 mol/kg H_2O) of chlorine are plotted in Fig. 4a and b to identify the effect of chlorine concentration on removing Pb and Cu. By comparing the results of Fig. 2a and b with Fig. 4a and b, we can clearly see that, when the concentration of chlorine rises from 1 to 10 mol/kg H_2O, the required acidity for conversions of PbS and Cu_2S to $PbCl_2$ and CuCl decreases, and meanwhile, the oxidation potential needed to dissolve $PbCl_2$ and CuCl also drops.

Impurities Removal from Low-Grade Molybdenite Concentrate

Effect of Hydrochloric Acid

The above results show that HCl can react with PbS, Cu_2S and FeS to dissolve impurities, meanwhile, it can affect the pH of leaching system, which is of great significance in removing impurities. So, in following experiments, the effect of hydrochloric acid concentration is studied firstly. The effect of HCl concentration on the removal of impurities and leaching ratio of molybdenum and rhenium were investigated under the conditions of keeping temperature at 95 °C, leaching time of 60 min and liquid–solid ratio of 4 (mL/g). The experimental results are shown in Fig. 5. As can be seen, with the HCl concentration increasing from 2 to 8 wt%, the removal ratio of Pb, Ca, Fe, and Cu sharply increases from 41.93, 67.18, 16.33, and 18.11%, to 91.75, 97.40, 31.21, and 24.32%, when the HCl concentration increasing to 20 wt% continually, the removal ratio almost remain constant. In addition, as the HCl concentration increases from 2 to 5 wt%, the loss ratio of molybdenum increases from 1.07% to 1.43%, while as the concentration reaches 8 wt%, the loss ratio sharply decreases to 0.87% and then almost remain constant. Moreover, when the HCl concentration increasing from 2 to 20 wt%, the loss ratio of rhenium increases from 2.35 to 5.15% continually. In conclusion, the above results clearly show that, when HCl concentration reaching 8 wt%, the removing ratio of impurities can reach a high level, while the loss ratio of molybdenum and rhenium remain relatively low, so the optimum concentration of HCl is 8 wt%.

Effect of Calcium Chloride

As mentioned above, raising the concentration of chlorine is conductive to removing impurities, but solely using HCl to enrich the concentration of chlorine would result in the severe corrosion of reactors. Herein, as calcium chloride is cheap, safe,

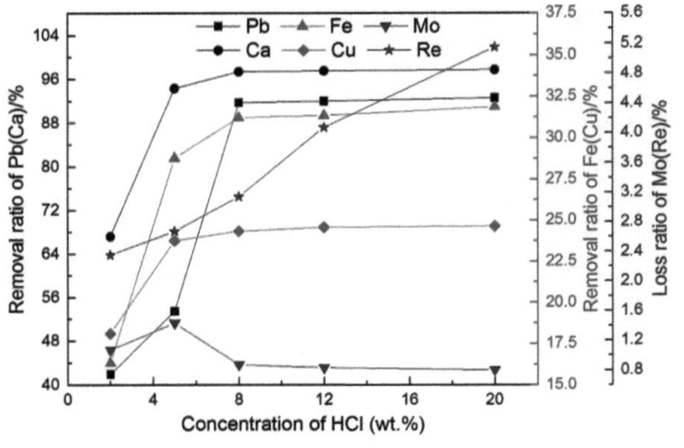

Fig. 5 The effect of HCl concentration on the leaching ratio of impurities, Mo and Re

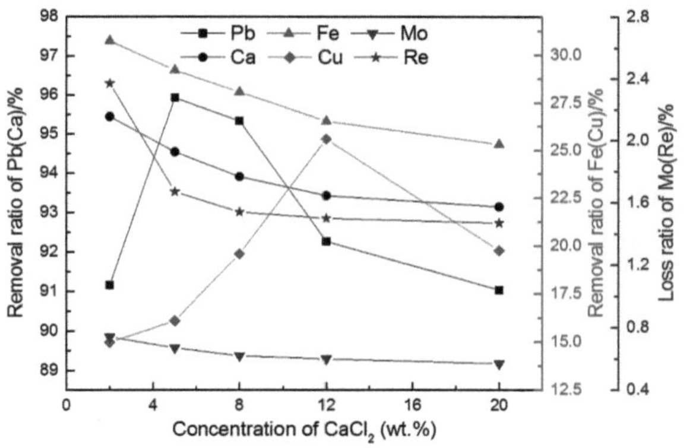

Fig. 6 The effect of $CaCl_2$ concentration on the leaching ratio of impurities, Mo and Re

easy to get and can provide more chlorine in the same condition, it is selected to removing impurities in following experiments. The effect of $CaCl_2$ concentration on the removal ratio of impurities and leaching ratio of molybdenum and rhenium were investigated under the conditions of keeping hydrochloric acid of 8 wt%, temperature at 95 °C, leaching time of 60 min and liquid–solid ratio of 4 (mL/g). The experimental results are shown in Fig. 6. It is clear that, the removal ratio of Pb and Cu increases to the optimum of 95.93 and 25.58%, as the $CaCl_2$ concentration increases to 5 and 12 wt%, respectively. The variation of Ca is similar to that of Fe, as the $CaCl_2$ concentration increases from 2 to 20 wt%, the removal ratio of Ca and Fe decreases from 95.45 and 30.74% to 93.15 and 25.30%, respectively. In addition, as the $CaCl_2$ concentration increases from 2 to 8 wt%, the loss ratio of molybdenum decreases from 0.74 to 0.62%, and when the concentration reaching 20 wt%, the loss ratio of molybdenum slightly decreases to 0.57%, while as for rhenium, when the $CaCl_2$ concentration increasing from 2 to 5 wt%, the loss ratio of rhenium sharply decreases from 2.37 to 1.67%, and when the $CaCl_2$ concentration increases to 20 wt%, the loss ratio of rhenium decreases to 1.47% slightly. Based on above results, we can clearly see that, the experimental results are basically consistent with the E-pH calculation results, raising the concentration of $CaCl_2$ within a certain range is beneficial to removing Pb and Cu. In addition, $CaCl_2$ behaves well in retaining molybdenum and rhenium.

Effect of $CuCl_2$ and $FeCl_3$ on Removal of Impurities

Effect of Copper Dichloride

Based on the above E-pH diagrams, we can clearly see that, the gangue minerals of PbS, Cu_2S and FeS can dissolve, by raising the oxidation potential within a suitable range of pH. So, the typical oxidizing chlorides of $CuCl_2$ and $FeCl_3$ are chosen to

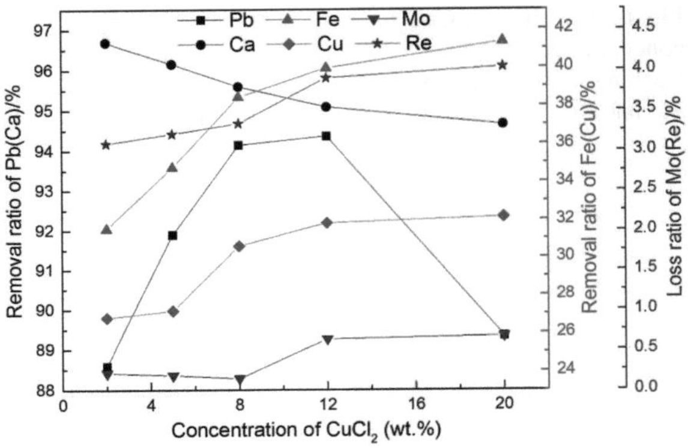

Fig. 7 The effect of $CuCl_2$ concentration on the leaching ratio of impurities, Mo and Re

remove impurities in following experiments. Fig. 7 is the experimental results of the effect of $CuCl_2$ concentration on the removal of impurities and leaching ratio of molybdenum and rhenium. The experiments were performed in the concentration of $CuCl_2$ range from 2 to 20 wt% under the conditions: HCl concentration of 8 wt%, temperature at 95 °C, leaching time at 60 min and liquid–solid ratio at 4 (mL/g). As can be seen, the removal ratio of Pb increases sharply from 88.58 to 94.14%, when $CuCl_2$ concentration increasing from 2 to 8 wt%, and then reaches the maximum of 94.36% in 12 wt%, while, when the $CuCl_2$ concentration increasing to 20 wt%, the removal ratio sharply decreases to 89.34%. The variation of Fe is similar to that of Cu, as the $CuCl_2$ concentration increases from 2 to 8 wt%, the removal ratio sharply increases from 31.50 and 26.78% to 38.43 and 30.95% respectively, when the $CuCl_2$ concentration increases continually, the removal ratio of Fe and Cu slightly increases. While as the $CuCl_2$ concentration increases from 2 to 20 wt%, the removal ratio of Ca gradually decreases from 96.69 to 94.65%. In addition, as the $CuCl_2$ concentration increasing from 2 to 8wt%, the loss ratio of Mo and Re remain at a relatively low level of 0.14 and 3.34%, while, when the concentration reaching 12 wt%, the removal ratio sharply increases to 0.63 and 3.90%, respectively, and then almost remain constant. From the above results, we can clearly see that, raising $CuCl_2$ concentration within a certain range is beneficial to removing Pb, Fe and Cu, while it may cause the large loss of rhenium.

Effect of Ferric Chloride

The effect of $FeCl_3$ concentration on the removal of impurities and leaching ratio of molybdenum and rhenium were investigated under the conditions of keeping hydrochloric acid concentration of 8 wt%, temperature at 95 °C, leaching time of 60 min and liquid–solid ratio of 4 (mL/g). The experimental results were plotted in Fig. 8. As can be seen, the variation of Pb is similar to that of Fe, the maximum removal ratio of 93.97% and 37.14% is reached for Pb and Fe, as the concentration of

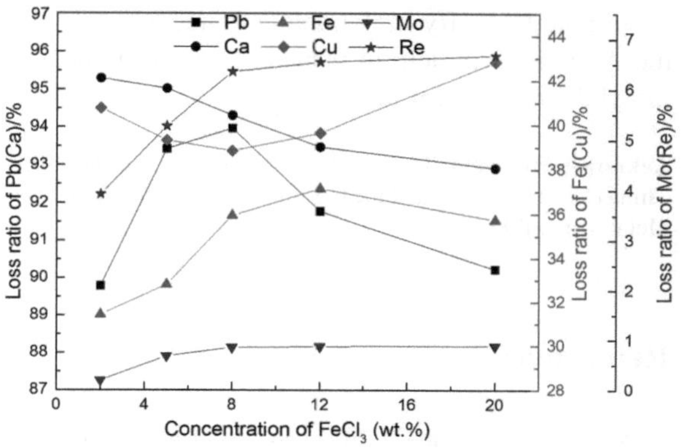

Fig. 8 The effect of $FeCl_3$ concentration on the leaching ratio of impurities, Mo and Re

$FeCl_3$ increases to 8 and 12 wt%, respectively, and then increasing the concentration continually, the removal ratio begins to decrease sharply. While as the $FeCl_3$ concentration increasing from 2 to 8 wt%, the removal ratio of Cu slowly decreases from 40.75 to 38.84%, but, when the $FeCl_3$ concentration increasing to 20 wt% continually, the removal ratio increases to 42.83%. The increasing of $FeCl_3$ concentration is not conducive to removing Ca, when the $FeCl_3$ concentration increasing from 2 to 20 wt%, the removing ratio of Ca gradually decreases from 95.21 to 92.92%. In addition, the variation of Mo is similar to that of Re, when the $FeCl_3$ concentration increasing from 2 to 8 wt%, the loss ratio of molybdenum and rhenium sharply increases from 0.19 and 3.92% to 0.85 and 6.35%, respectively, and when the concentration increasing to 20 wt%, the loss ratio almost remaining constant. According to above results, we can clearly see that, raising the concentration of $FeCl_3$ in a certain range is beneficial to removing Pb, Fe and Cu, but it may cause a large loss of molybdenum and rhenium.

Conclusions

In this study, E-pH calculations were used to analyze the deep removal of impurities from low-grade molybdenite concentrates. The results indicate that acid system ($pH < 2$) is beneficial to the removal of impurities and the reservation of molybdenum and rhenium. When the pH is around zero, the oxidation potential for gangue minerals to dissolve follows the order: E (Cu_2S) > E (PbS) > E (FeS). The mixture of HCl and $CaCl_2$ is superior in removing Pb, Cu, and retaining Mo, Re, and thus suitable to work as purification agent for rhenium-bearing molybdenite concentrates with high-content lead and copper. In addition, the experimental results also show that, adding oxidative chloride ($CuCl_2$, $FeCl_3$) into hydrochloric acid is beneficial to the removal of most impurities (Pb, Fe, Cu), while it may cause a large loss of rhenium. Therefore, the mixture of hydrochloric acid and oxidative chloride ($CuCl_2$, $FeCl_3$)

is suitable to purify molybdenite concentrates without rhenium. In conclusion, all these studies can help to provide targeted guidance in purifying different types of molybdenite concentrate.

Acknowledgements This work was supported by the National Natural Science Foundation of China (No. 51874355) and the Co-Innovation Center for Clean and Efficient Utilization of Strategic Metal Mineral Resources.

References

1. Xia Y, Xiao L, Xiao C (2015) Direct solvent extraction of molybdenum(VI) from sulfuric acid leach solutions using PC-88A. Hydrometallurgy 158:114–118
2. Henckens MLCM, DriessenPPJ Worrell E (2018) Molybdenum resources: their depletion and safeguarding for future generations. Resour Conserv Recycl 134:61–69
3. Ammann PR, Loose TA (1972) Rhenium volatilization during molybdenite roasting. Metall Trans 3(4):1020–1022
4. Juneja JM, SinghS Bose DK (1996) Investigations on the extraction of molybdenum and rhenium values from low grade molybdenite concentrate. Hydrometallurgy 41(2–3):201–209
5. Li G, You Z, Sun H (2016) Separation of rhenium from lead-rich molybdenite concentrate via hydrochloric acid leaching followed by oxidative roasting. Metals—Open Access Metall J 6(11):282
6. Austin JW, Bradburn RG, Cromwell CA (1985) Process for purifying molybdenite concentrates. US. Patent 450,049. 6 Feb 1985

Production of High-Purity Titanium Dioxide from Spent Selective Catalytic Reduction (SCR) Catalyst

Gyeonghye Moon, Jin-Hyung Kim, Yeon-Chul Cho, In-hyeok Choi, Hee-Nam Kang, Tae-Hyuk Lee, Jin-Young Lee and Jungshin Kang

Abstract In order to produce high-purity TiO_2 using spent SCR catalyst, a novel hydrometallurgical process was investigated. In water leaching, the influence of temperature, basicity, and solid/liquid ratio on dissolution of product (mainly, sodium titanate) obtained by soda-melting of catalyst at 1273 K was investigated. In addition, the influence of HCl concentration and temperature on HCl leaching was investigated to increase the leaching efficiency by retarding of TiO_2 generation Silicon in titanium solution, obtained after HCl leaching, was precipitated during hydrolysis. Therefore, for the removal of silicon in titanium solution, optimal conditions to decrease the silicon concentration below 1 ppm were determined. When precipitates obtained by hydrolysis of purified titanium solution at 363 K were calcined at 973 K, 99.6–99.9% TiO_2 was obtained. Therefore, this study demonstrates the feasibility of this cost-effective method for production of high-purity TiO_2 from spent SCR catalyst.

Keywords Spent SCR catalyst · Titanium dioxide · Pigment · Leaching
Silicon removal · Hydrolysis

Introduction

The environmental concerns regarding air pollution by nitrogen oxides (NO_x) gas generated from stationary and mobile applications have recently increased. To prevent the emissions of NO_x gas, a selective catalytic reduction (SCR) catalyst is used in power plants and automobiles. When an exhaust gas including NO_x passes through the SCR catalyst chamber, NO_x gas is reduced to nitrogen (N_2) gas by reacting

G. Moon · J-H. Kim · Y-C. Cho · H-N. Kang · T-H. Lee · J. Lee · J. Kang (✉)
Korea Institute of Geoscience and Mineral Resources, 124 Gwahak-ro, Yuseong-gu, Daejeon 34132, Korea
e-mail: jskang@kigam.re.kr

I. Choi · J. Lee · J. Kang
Department of Resources Recycling, University of Science and Technology, 217 Gajeong-ro, Yuseong-gu, Daejeon 34113, Korea

G. Azimi et al. (eds.), *Rare Metal Technology 2019*, The Minerals, Metals & Materials Series, https://doi.org/10.1007/978-3-030-05740-4_13

with ammonia (NH_3) [1]. Therefore, with an increase in environmental concerns, the volume of the SCR catalysts installed has gradually increased in South Korea, particularly in power plants. For example, the volume of the SCR catalyst installed in power plants increased from 18,752 m^3 in 2008 to 27,455 m^3 in 2012 [2].

Generally, the SCR catalyst consists of vanadium pentoxide (V_2O_5, 0.5–1.5 wt%) for the catalytic agent, tungsten trioxide (WO_3, 7–10 wt%) for the promoter, and titanium dioxide (TiO_2, 70–80 wt%) for the supporting materials [3–5]. After a SCR catalyst has been used for approximately 3 years after its initial installation, regeneration of the spent catalyst is carried out to restore the degraded catalytic performance [3, 5]. Although regeneration processing is conducted, the decline of the catalytic performance is inevitable owing to a degradation of the texture, porosity, and structure of the SCR catalyst [6]. As a result, a spent SCR catalyst that cannot be further regenerated is discarded in a landfill. Therefore, when the increase in the amount of spent SCR catalysts generated is taken into consideration, the recovery of valuable elements such as vanadium (V), tungsten (W), and titanium (Ti) will become an important issue in the near future.

Although the importance of the recycling of spent SCR catalysts has increased, only a few investigations into the recovery of V, W, and Ti from the spent SCR catalyst have been carried out [5–9]. In addition, most studies have concentrated on the recovery of V and W from spent SCR catalysts utilizing hydrometallurgical methods, despite 70–80% of TiO_2 being used as a supporting material. Although only a few studies have been conducted on spent SCR catalysts, the recovery methods for V, W, or Ti from primary or secondary resources such as desulfurization catalysts or titania slag have been developed by many researchers [10–13]. A pressure leaching or roast-leaching process using an alkaline medium such as sodium hydroxide (NaOH) or sodium carbonate (Na_2CO_3) has been reported for the recovery of W [10]. In addition, an aqueous alkaline medium has been used for the recovery of V from slags, ashes, and catalysts [11, 12]. Furthermore, the production of TiO_2 from titania slag through alkaline roasting, leaching, and solvent extraction has been investigated [13].

In this study, a novel and cost-effective hydrometallurgical process was investigated for the production of high-purity TiO_2 using a spent SCR catalyst, as shown in Fig. 1. High-purity TiO_2 was produced through soda-melting of the spent SCR catalyst using Na_2CO_3, water leaching of sodium titanate ($NaTiO_x$), concentrated hydrochloric acid (HCl) leaching of the residues obtained after the water leaching, the removal of silicon (Si) in the Ti solution obtained after the HCl leaching, and hydrolysis of the purified Ti solution. The results of this study demonstrate the feasibility of the production of high-purity TiO_2 from a spent SCR catalyst using a cost-effective hydrometallurgical method.

Experimental

Table 1 shows the analytical results of a spent SCR catalyst feed and $NaTiO_x$ obtained after soda-melting of the catalyst. During the soda-melting step, 40 kg of a spent SCR

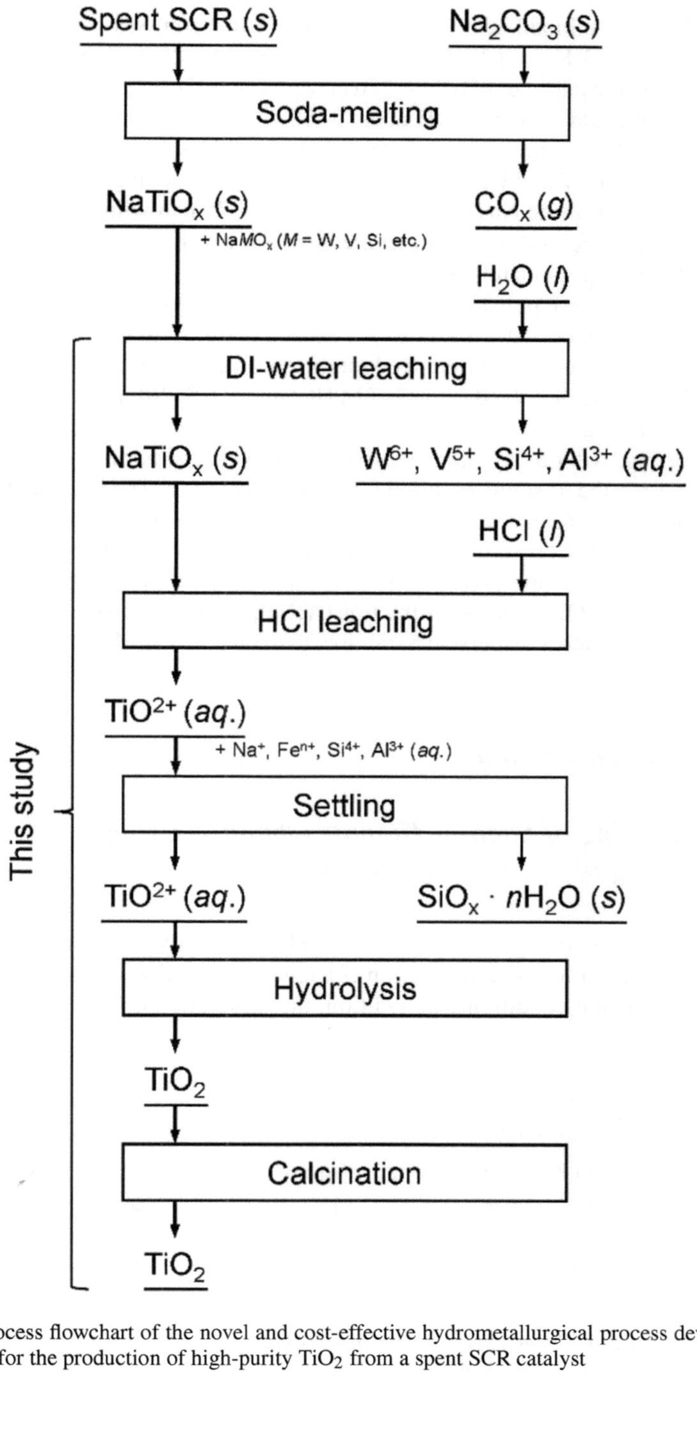

Fig. 1 Process flowchart of the novel and cost-effective hydrometallurgical process developed in this study for the production of high-purity TiO_2 from a spent SCR catalyst

catalyst from a combined heat and power plant in Incheon was reacted with 48 kg of Na_2CO_3 at 1273 K for 1 h. The $NaTiO_x$ obtained by the soda-melting step was pulverized, and the particles of below 45 μm were leached using deionized (DI) water with/without the addition of NaOH under 2–20% solid/liquid (S/L) ratio (weight of feedstock (g) x 100/volume of solution (ml)) at 303–373 K for 1 h in a Teflon reactor for the extraction of V and W. The feedstock for the HCl leaching was prepared using DI water leaching of the $NaTiO_x$ particles of below 75 μm without the addition of NaOH under 20% S/L ratio at room temperature for 1 h. To investigate the influence of the HCl concentration and temperature on the efficiency of the Ti dissolution, the residues obtained were pulverized, and the particles of below 75 μm were dissolved using 3–7 M HCl under 10% S/L ratio at 333–353 K for 3 h to produce a Ti solution. To prepare the Ti solution used as a feed for the removal of Si, the $NaTiO_x$ particles of below 105 μm were dissolved using 5–7 M HCl under 10% S/L ratio at 333 K for 3 h. This Ti solution was contained in a polyethylene bottle and preserved at 274–313 K without stirring. After 5 days had elapsed, centrifugal separation and filtration were conducted using the Ti solution stored for the removal of Si. The purified Ti solution was hydrolyzed at 363 K for 5 h and calcined at 973 K for 1 h to obtain TiO_2.

The compositions, crystalline phases, and surfaces of the samples obtained during the experiments were analyzed using inductively coupled plasma optical emission spectroscopy (ICP-OES), X-ray diffraction (XRD), and field emission scanning electron microscopy/energy dispersive X-ray spectroscopy (FE-SEM/EDS), respectively.

Results and Discussion

Water Leaching of Sodium Titanate Obtained Using Soda-Melting of Spent SCR Catalyst

Table 2 shows the experimental conditions and fractions of V, W, Al, and Si dissolved using DI water leaching. The dissolution efficiency of V and W was larger than 99.2% for all experimental conditions. In addition, the dissolution efficiency of Al and Si

Table 1 Analytical results of the spent SCR catalyst feed and residues obtained after soda-melting

	Concentration of element i, C_i (mass%)							
	TiO_2[aa]	WO_3[a]	V_2O_5[a]	Al_2O_3[a]	Fe_2O_3[a]	CaO[a]	SiO_2[b]	Na_2O[a]
Feed (spent SCR catalyst)	67.2	7.49	3.23	1.59	0.33	1.80	7.15	N.A
Residues after soda-melting	34.3	4.35	1.88	1.23	0.36	0.77	4.71	39.3

[a]Determined by ICP-OES analysis, N.A: Not Analyzed
[b]Determined by wet analysis (gravimetric analysis)

Fig. 2 Results of FE-SEM/EDS analysis for the residues obtained after DI water leaching

was increased when the reaction temperature or basicity was increased, or the S/L ratio was decreased. In addition, the amount of Ti dissolved from the soda-melted spent SCR catalyst through DI water leaching was negligible when the leachate was analyzed using ICP-OES.

The compounds in the residues obtained after the DI water leaching were not identified through an XRD analysis. However, as shown in Fig. 2, when the residues were analyzed using FE-SEM/EDS, the main elements identified were Na, Ti, and O. Therefore, it is expected that the main compound in the residues was $NaTiO_x$.

Table 2 Experimental conditions and fraction of each element dissolved through DI water leaching

Exp. no.[a]	Feed weight, w_{melted_scr}/g	NaOH weight, w_{NaOH}/g	Temp, T/K	S/L ratio (%)[c]	Fraction of each element dissolved, X_M (%)[b]			
					Al	V	W	Si
180108-1	10	–	303	2	61.9	99.3	99.3	80.2
180514	10	–	343	2	85.3	99.3	99.2	84.8
180201-1	10	15	343	2	98.6	99.7	99.6	93.7
180223-1	100	–	343	20	23.4	99.7	99.5	73.5

[a]Experimental conditions: The Teflon reactor was heated using mantle heater, reaction time = 1 h, agitation = 500 rpm, particle size = under 45 μm
[b]Fraction of each element dissolved = 100 × (1 − (weight of element in residue/weight of element in feedstock)); concentration of all elements except Si was analyzed using ICP-OES; concentration of Si was analyzed using wet analysis (gravimetric analysis)
[c]S/L ratio = weight of feedstock (g) x 100/volume of solution (ml)

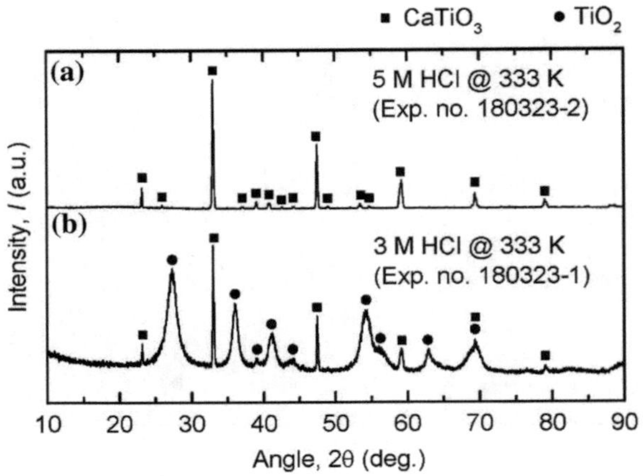

Fig. 3 Results of XRD analysis of the residues obtained after the HCl leaching

HCl Leaching of Sodium Titanate Obtained After DI-Water Leaching

Table 3 shows the experimental conditions and the fractions of Na, Ti, Fe, and Si dissolved using HCl leaching. As shown in Table 3, when the $NaTiO_x$ obtained after the water leaching was dissolved using HCl at 333 K for 3 h, the dissolution efficiency of Ti increased from 29.6 to 96.5% by increasing the concentration of HCl used from 3 to 5 M. Figure 3 shows the results of an XRD analysis of the residues obtained after the HCl leaching. When the $NaTiO_x$ was dissolved using 3 M HCl, TiO_2 was identified along with calcium titanate ($CaTiO_3$). However, no TiO_2 was identified when 5 M HCl was used for the dissolution of $NaTiO_x$. Therefore, to increase the dissolution efficiency of Ti by retarding the generation of TiO_2 during the HCl leaching under a certain reaction temperature, a sufficient concentration of HCl is required depending on the HCl leaching conditions. In addition, as shown in Table 3, when the reaction temperature of the HCl leaching was increased from 333 to 353 K, the concentration of HCl was increased from 5 to 7 M to obtain a high dissolution efficiency of Ti.

Removal of Si from Ti Solution Produced through HCl Leaching

During the HCl leaching, a certain amount of Si was contained in the Ti solution depending on the leaching conditions, as shown in Table 3. The Si in the Ti solution affects the filtration of the solution and/or the purity of the TiO_2 obtained after hydrolysis of the Ti solution depending on the experimental conditions. Therefore,

Table 3 Experimental conditions and the fraction of each element dissolved through HCl leaching

Exp. no.[a]	Feed weight, $w_{water\_leached}$/g	Concentration of HCl (M)	Temp, T/K	S/L ratio (%)[c]	Fraction of each element dissolved, X_M (%)[b]			
					Na	Ti	Fe	Si
180323-1	25.0	3	333	10	95.8	29.6	92.4	70.1
180323-2	25.0	5	333	10	99.2	96.5	99.2	99.0
180703	25.0	6	343	10	99.6	96.8	99.4	97.4
180325-2	25.0	7	353	10	99.2	96.2	99.7	44.9

[a]Experimental conditions: Double jacket reactor was heated by bath circulator, reaction time = 3 h, agitation = 500 rpm, particle size = under 75 μm
[b]Fraction of each element dissolved = 100 × (1 − (weight of element in residues/weight of element in feedstock))
[c]S/L ratio = weight of feedstock (g) x 100/volume of solution (ml)

Table 4 Experimental conditions and concentration of Si in Ti solution after 5 days have elapsed

Concentration of HCl used for HCl leaching (M)	Temperature for preservation, *T*/K	Concentration of Si in the Ti solution (mg/L)[a]
Ti solution feed (5 M)		1114
5	274	1098
	313	N.D

N.D not detected (<1 ppm)
[a]The concentration of Si was analyzed using ICP-OES

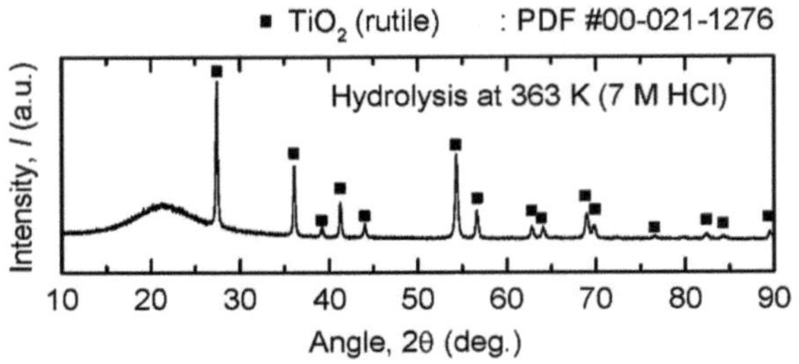

Fig. 4 Results of XRD analysis of the product obtained after hydrolysis and calcination

the removal of Si from the Ti solution is required to increase the productivity of the processing and the purity of the product. Table 4 shows the experimental conditions and concentration of Si in the Ti solution after 5 days have elapsed. As shown in Table 4, when the Ti solution produced through HCl leaching using 5 M HCl was preserved at 274 K, Si in the Ti solution was rarely removed. However, when the Ti solution was preserved at 313 K, the concentration of Si in the Ti solution was below the detection limit of ICP-OES. It is expected that the silicic acid in the Ti solution was polymerized and sufficiently grown to silica gel for filtration [14].

Hydrolysis of Purified Ti Solution Obtained After the Removal of Si

Table 5 shows the analytical results of the product obtained after hydrolysis of the purified Ti solution and calcination. In addition, Figs. 4 and 5 show the results of the XRD analysis and image of the product obtained, respectively, after hydrolysis and calcination. As shown in Table 5 and Fig. 4, 99.6–99.9% of the TiO_2 powder (synthetic rutile) was obtained.

Table 5 Analytical results of TiO_2 obtained after hydrolysis and calcination

Concentration of HCl used for HCl leaching	Concentration of element i, C_i (mass%)[a]								Note[b]
	Al_2O_3	Na_2O	Fe_2O_3	CaO	WO_3	V_2O_5	SiO_2	TiO_2	
5 M HCl	N.D	N.D	N.D	N.D	N.D	N.D	N.D	Bal.	$TiO_2 \geq 99.9\%$
7 M HCl	N.D	N.D	N.D	N.D	0.43	N.D	N.D	Bal.	$TiO_2 = 99.57\%$

[a]Concentration was determined using ICP-OES analysis, N.D: Not Detected (< 0.010 mass%)

[b]Concentration of TiO_2 was calculated as follows: TiO_2 purity = 100 − (conc. of oxides of Al, Na, Fe, Ca, W, V, and Si)

Fig. 5 Results of FE-SEM analysis of the product obtained after hydrolysis and calcination

Conclusion

A novel and cost-effective hydrometallurgical process was investigated for the production of high-purity TiO_2 from a spent SCR catalyst. When sodium titanate particles of below 45 μm were leached using DI water under 2–20% S/L ratio at 303–373 K for 1 h, more than 99.2% of the V and W were extracted. During the HCl leaching step, above 96.2% of Ti was extracted from the residues obtained after the water leaching when particles of the feedstock of below 75 μm were dissolved using 5–7 M HCl under 10% S/L ratio at 333–353 K for 3 h. In addition, the concentration of Si in the Ti solution was below 1 ppm by controlling the temperature for the preservation. When the purified Ti solution was hydrolyzed at 363 K and calcined at 973 K, 99.6–99.9% TiO_2 was obtained.

Acknowledgements This research was supported by the R&D center for valuable recycling (Global-Top Environmental Technology Development Program) funded by the Korean Ministry of the Environment in Korea (Project No.: GT-11-C-01-230-0).

References

1. BuscaG LiettiL, Ramis G, Berti F (1998) Chemical and mechanistic aspects of the selective catalytic reduction of NO_x by ammonia over oxide catalysts: a review. Appl Catal B: Environ 18:1–36
2. Kim H (2012) Recovery of valuable metals from spent SCR catalysts for de-NO_x by soda roasting and solvent extraction. Korea University of Science and Technology

3. Spivey JJ (2002) Catalysis, vol 16. The Royal Society of Chemistry, North Yorkshire
4. Marberger A, Elsener M, Ferri D, Krocher O (2015) VOx surface coverage optimization of V_2O_5/WO_3-TiO_2 SCR catalysts by variation of the V loading and by aging. Catalysts 5:1704–1720
5. Choi I, Kim H, Moon G, Jyothi RK, Lee J (2018) Spent V_2O_5-WO_3/TiO_2 catalyst processing for valuable metals by soda roasting-water leaching. Hydrometallurgy 175:292–299
6. Choi I, Moon G, Lee J, Jyothi RK (2018) Hydrometallurgical processing of spent selective catalytic reduction (SCR) catalyst for recovery of tungsten. Hydrometallurgy 178:137–145
7. Wu W-C, Tsai T-Y, Shen Y-H (2016) Tungsten recovery from spent SCR catalyst using alkaline leaching and ion exchange. Minerals 6(107):1–10
8. Choi I, Moon G, Lee J, Jyothi RK (2018) Extraction of tungsten and vanadium from spent selective catalytic reduction catalyst for stationary application by pressure leaching process. J Clean Prod 197:163–169
9. Kim JW, Lee WG, Hwang IS, Lee JY, Han C (2015) Recovery of tungsten from spent selective catalytic reduction catalysts by pressure leaching. J Ind Eng Chem 28:73–77
10. Gaur RPS (2006) Modern hydrometallurgical production methods for tungsten. JOM 58:45–49
11. Vitolo S, Seggiani M, Falaschi F (2001) Recovery of vanadium from a previously burned heavy oil fly ash. Hydrometallurgy 62:145–150
12. Navarro R, Guzman J, Saucedo I, Revilla J, Guibal E (2007) Vanadium recovery from oil fly ash by leaching, precipitation and solvent extraction processes. Waste Manag 27:425–438
13. Middlemas S, Fang ZZ, Fan P (2013) A new method for production of titanium dioxide pigment. Hydrometallurgy 131–132:107–113
14. Lefler H, Zhang Y, Fang ZZ, Free M, Huang Z (2017) Removal of silicon from highly acidic HCl medium to produce purified TiO_2. Hydrometallurgy 173:218–223

Reduction of $TiCl_4$ to TiH_2 with CaH_2 in Presence of Ni Powder

Mohammad Rezaei Ardani, Aws Sadoon Mohammed Al Janabi, Sanjith Udayakumar, Sheikh Abdul Rezan Sheikh Abdul Hamid, Abdul Rahman Mohamed, H. L. Lee and Ismail Ibrahim

Abstract Titanium hydride (TiH_2) is of significant interest for titanium powder metallurgy and hydrogen storage material. TiH_2 can be produced from titanium tetrachloride ($TiCl_4$) reduced with calcium hydride (CaH_2). In this study, the effects of temperature, time and amount of Ni as catalyst on the reduction process were evaluated. Temperature was between 300 and 500 °C, reaction time was set between 3 and 5 h, and weight percentage of Ni to the CaH_2 was from 10 to 50%. The weight of final powders was measured by element hydrogen analysis after each experiment. The highest weight gain (89.11%) was obtained for the sample reduced at 500 °C with 50% Ni catalyst for 3 h. Samples with higher weight gain were characterized by X-ray diffraction (XRD) and scanning electron microscope (SEM). The results indicate that producing of TiH_2 from reduction of $TiCl_4$ with CaH_2 in the presence of Ni was feasible and could lead to low-temperature synthesis of TiH_2.

Keywords Powder metallurgy · Titanium · Titanium hydride
Design of experiment

M. R. Ardani · A. S. M. Al Janabi · S. Udayakumar · S. A. R. Sheikh Abdul Hamid (✉)
School of Materials & Mineral Resources Engineering, Engineering Campus, Universiti Sains Malaysia, 14300 Nibong Tebal, Penang, Malaysia
e-mail: srsheikh@usm.my

I. Ibrahim
Mineral Research Centre, Mineral and Geoscience Department Malaysia, Jalan Sultan Azlan Shah, 31400 Ipoh, Perak, Malaysia
e-mail: is.ibrahim@jmg.gov.my

A. R. Mohamed
School of Chemical Engineering, Engineering Campus, Universiti Sains Malaysia, 14300 Nibong Tebal, Pulau Pinang, Malaysia

H. L. Lee
School of Chemical Sciences, Universiti Sains Malaysia, 11800 USM Penang, Malaysia

G. Azimi et al. (eds.), *Rare Metal Technology 2019*, The Minerals, Metals & Materials Series, https://doi.org/10.1007/978-3-030-05740-4_14

Introduction

Titanium (Ti) and its alloys have unique properties such as low density, high temperature capabilities, high ratio of strength to weight, outstanding resistance to corrosion and high toughness in aggressive environment which made them to be widely used for structural purposes, power generation, and chemical, medical, and sports industries. Furthermore, titanium alloys have gained widespread application in aerospace industry owing to its resistance to severe operating conditions including high temperature, high pressure and hydrogen rich environment [1–4].

One of the main industrial methods of titanium production is the Kroll process. This method is based on the extraction of titanium from TiO_2 in an expensive batch process. The TiO_2 is converted to $TiCl_4$ in a chlorination process and then the titanium sponge is produced via the subsequent reduction of $TiCl_4$ with Mg. In addition to the high production costs, the manufacturing process is highly increased due to titanium hard workability [5]. Today the production of titanium is mainly based on powder metallurgy. While the cost of titanium production by traditional ingot technology is significantly high, titanium powder is of a great attention due to its production costs and tremendous chemical and physical properties of titanium powder especially in biomaterials [6, 7], additive manufacturing [8] and aerospace [9]. Ti powder is produced either by gas atomization (spherical, >10 μm particle size) or via the hydride-dehydride method. In order to increase the sintered density and balance of properties of titanium alloys without additional thermomechanical operations, the titanium hydride (TiH_2) powder instead of conventional titanium powder can be successfully employed, as starting material [10–12]. Moreover, increase in densification was reported using TiH_2 due to reduction in cold welding and friction phenomena occurring during the compaction process [13–15].

In this study, a laboratory scale study was performed to develop a low cost and low temperature process for synthesizing TiH_2 powder by means of reduction of $TiCl_4$ with CaH_2.

Experimental

TiH_2 is produced by the gas-solid hydriding reaction between CaH_2 and $TiCl_4$. The expected chemical reaction is as Eq. 1:

$$TiCl_4(g) + CaH_2 = TiH_2 + 2CaCl_2 + H_2(g) \quad \Delta G^\circ = -589.94 - 0.0763T \text{ (kJ/mol)} \tag{1}$$

The value for standard Gibbs free energy change (ΔG°) indicated that the reaction was thermodynamically feasible and spontaneous at lower temperatures (<350 °C). The reaction between CaH_2 and $TiCl_4$ was performed in a setup illustrated in Fig. 1. H_2 gas was passed through the phosphorous pentoxide to absorb any moisture in the hydrogen gas. Then, it was allowed to enter into the $TiCl_4$ bottle. The flow rate of hydrogen gas was monitored and controlled regularly during the experiment. The

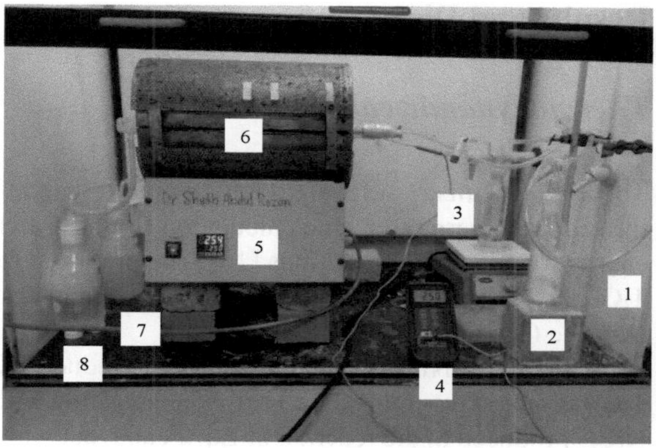

Fig. 1 Experimental Setup. (From the right, the hydrogen gas passed through the phosphorous pentoxide and ended up at $TiCl_4$ scrubber). 1—H_2, 2—Phosphorus Pentoxide, 3—Titanium Tetrachloride, 4—Thermocouple, 5—Furnace, 6—Sample in Furnace, 7—Sodium Hydroxide, 8—Hydrochloric Acid

$TiCl_4$ was heated up to 140 °C in order to turn to gaseous state and the flowing of H_2 gas was used to carry the $TiCl_4$ gas through the quartz tube in the furnace. The CaH_2 pellet was placed in a crucible with dimension of (L: 63.5 mm, W: 16.5 mm, D: 7.3 mm) at the center of the quartz tube at the desired temperature. Residual exit gas from the tube was neutralized with the hydrochloric acid (HCl) and NaOH before excess gas was released to the atmosphere. The Ni catalyst was used in this study. Three different ratios between Ni and CaH_2 were used. The ratios were 1:10, 3:10, and 5:10. These three ratios were presented in the study as 10, 30, and 50% of the catalyst.

Design of Experiment (DOE), a statistical approach was used for this study with the help of Design-Expert version 11 software (State-Ease, USA). The study was carried out by varying three different factors for three distinct levels namely; temperature was between 300 and 500 °C, reaction time was set between 3 and 5 h, and weight percentage of Ni to the CaH_2 was from 10 to 50%. The parameters were chosen based on the dehydriding temperature of CaH_2. DOE was used to study the interaction parameters with the measured responses so that the most significant parameters in the reaction can be identified. The phase compositions and morphology of the reaction products as well as the extent of dehydriding in the reaction were studied by X-ray diffraction using Cu-Kα radiation with the wavelength of $\lambda = 1.5404$ Å (XRD; Bruker D8-advance, USA). The surface morphology and the elemental composition of the reaction product were studied by scanning electron microscopy equipped with electron dispersive X-ray spectroscopy (SEM/EDX; Leo Supra 35VP, Germany). The degree of dehydriding was measured using elemental CHNS/O analyzer (Perkin Elmer Series II model 2400, USA) by determining the hydrogen content present in the reactants before and after the dehydriding process.

Results and Discussion

Thermodynamic calculations

The HSC Chemistry software (Outokumpu Research Oy, Finland) was used to make the required thermodynamic calculation to find the changes in standard Gibbs free energy for the studied reactions [16]. The predicted reaction to synthesis TiH_2 from $TiCl_4$ in the present study was to react calcium hydride (CaH_2) with titanium tetrachloride ($TiCl_4$) as Eq. 1.

In this study, Ni was added as catalyst to reduce $TiCl_4$ to $TiCl_3$ as shown in Eq. 2. $TiCl_3$ is more favourable than $TiCl_4$ when reacting with CaH_2 to from TiH_2. Also, the $NiCl_2$ produced in the reaction of Eq. 2 can react with CaH_2 as Eq. 3 creating nickel monohydride, which will form TiH_2 as shown in Eq. 4. $NiCl_2$ from the previous reaction can recycle and repeat the reactions in Eqs. 3 and 4.

$$Ni + 2TiCl_4(g) = NiCl_2 + 2TiCl_3 \quad \Delta G^\circ = -122.2 + 0.331T \text{ (kJ/mol)} \tag{2}$$

$$2CaH_2 + 2NiCl_2 = 2CaCl_2 + H_2 + 2NiH(g) \tag{3}$$

$$2NiH + TiCl_4(g) = TiH_2 + 2NiCl_2 \tag{4}$$

The thermodynamic evaluation of the reaction confirms the feasibility for reduction of gaseous $TiCl_4$ to $TiCl_3$ by Ni from 0 to 360 °C as the Eq. 2 has negative ΔG° value. The $TiCl_3$ would help to reduce CaH_2 to TiH_2. Indeed, $TiCl_3$ is a more favourable route than $TiCl_4(g)$ due to its lower ΔG° as shown in Eq. 5 compared with $TiCl_4(g)$ in Eq. 1. However, this benefit of Ni stops above 360 °C as ΔG° is zero and moving towards positive. High amount of Ni can react with TiH_2 creating $NiTiO_3$ phase as Eq. 6 (from 0 to 500 °C, $\Delta G^\circ \approx -350$ kJ/mol).

$$2TiCl_3 + 3CaH_2 = 2TiH_2 + 3CaCl_2 + H_2(g) \quad \Delta G^\circ = -730.83 - 0.112T \text{ (kJ/mol)} \tag{5}$$

$$Ni + TiH_2 + 3H_2O(g) = NiTiO_3 + 4H_2(g) \quad \Delta G^\circ = -325.58 - 0.013T \text{ (kJ/mol)} \tag{6}$$

On the other hand, Ni has no favorable route to react with TiH_2 nor $TiCl_4$ to produces $NiTi_2$ phase as shown in Eqs. 7 and 8 at the studied temperature (e.g. 300–500 °C) due to their positive value of ΔG°. Indeed, Ni cannot react with TiO_2 in this temperature as shown in Eq. 9 because of the high value of ΔG° (above 900 kJ/mol, from 0 to 500 °C). From the reactant side, Eq. 10 gives positive ΔG° value and proves that Ni cannot react with CaH_2 to form $2.941NiH_{0.68}$ phase from 0 to 500 °C. For these reasons, Ni may be considered as a good catalyst for this system. However, thermodynamic calculations show that moisture and oxygen leakage into the sample can form $Ni(TiO_3)$ phase as Eq. 11 and Eq. 12 (from 0 to 500 °C, $\Delta G^\circ \approx -350$ kJ/mol).

$$2TiH_2 + Ni = NiTi_2 + 2H_2(g) \quad \Delta G^\circ = 139.1 - 0.2694T \text{ (kJ/mol)} \tag{7}$$

$$2TiCl_4(g) + Ni + 4H_2(g) = NiTi_2 + 8HCl(g) \quad \Delta G^\circ = 618.34 - 0.299T \text{ (kJ/mol)} \tag{8}$$

$$2TiO_2 + Ni = NiTi_2 + 2O_2(g) \quad \Delta G^\circ = 1025.3 - 0.2112T\ \text{(kJ/mol)} \tag{9}$$

$$2.941Ni + CaH_2 = Ca + 2.941NiH_{0.68} \quad \Delta G^\circ = 161.78 - 0.020T\ \text{(kJ/mol)} \tag{10}$$

$$Ni + TiH_2 + 3/2O_2(g) = NiTiO_3 + H_2(g) \quad \Delta G^\circ = -325.58 - 0.014T\ \text{(kJ/mol)} \tag{11}$$

$$Ni + TiO_2 + H_2O(g) = NiTiO_3 + H_2(g) \quad \Delta G^\circ = -0.8521 + 0.044T\ \text{(kJ/mol)} \tag{12}$$

The equilibrium diagram can show that the molar ratio of $TiCl_4$ to CaH_2 can make a noticeable impact to the formation of TiH_2. The equilibrium diagrams were developed with different molar ratios between $TiCl_4$ and CaH_2 using HSC software. The equilibrium diagram is calculated using the GIBBS or SOLGASMIX solvers (Outokumpu Research Oy, Finland), which use the Gibbs energy minimization method [16]. Figure 2a shows 1:2 molar ratio between $TiCl_4$ and CaH_2 respectively at stoichiometric reaction ($TiCl_4(g) + 2CaH_2 = TiH_2 + 2CaCl_2 + H_2(g)$). As shown in Fig. 2a, 2 kmol of CaH_2 was reduced to TiH_2 in the temperature range from 0 to 500 °C. This means the $TiCl_4(g)$ amount should be 22.414 ml for each 2 mol of CaH_2 as stoichiometry of Eq. 1 to get a full reduction. The number of moles used in the studied system was 0.00713 mol of CaH_2, which resulted in using approximately 80 ml of $TiCl_4$.

This favourable ratio 1:2 may change under some circumstances and the flow rate of the $TiCl_4$ may increase, as the flow rate of mixture of $TiCl_4$ and H_2 gases is not completely under control. A new equilibrium diagram plotted to show the impact of the increment of $TiCl_4$ to a ratio of 2:2 with CaH_2, twice the amount required compared to the stoichiometry Eq. 1. In Fig. 2b, the equilibrium diagram shows the

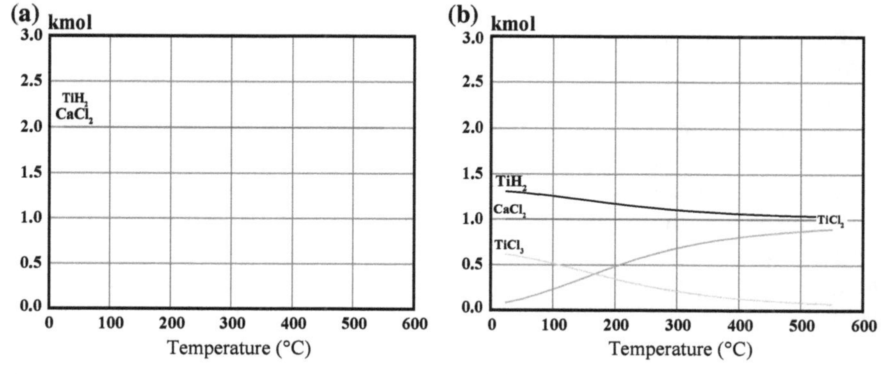

Fig. 2 **a** Equilibrium diagram shows the formation of TiH_2 if the ratio between $TiCl_4$ and CaH_2 is 1:2 mol. **b** Equilibrium diagram shows the formation of TiH_2, $CaCl_2$, $TiCl_2$, $TiCl_3$ if the ratio between $TiCl_4$ and CaH_2 is increased to 2:2 mol

formation of TiH_2 with a molar ratio of 2:2 between $TiCl_4$ and CaH_2, respectively. The increment of the $TiCl_4$ gas produces new phases in the system such as $TiCl_2$, $TiCl_3$, and $CaCl_2$. While, there is some decrement follows the line of TiH_2 formation from 100 to 400 °C; however, the formation of TiH_2 is still possible. In summary, the increment of $TiCl_4$ gas in the system by doubling the amount compared to the stoichiometry ratio may impacts the formation of TiH_2 negatively. Overall, from equilibrium diagrams, the best ratio of $TiCl_4$ and CaH_2 is 1:2 as stoichiometry ratio. However, the high amount of $TiCl_4$ produces new phases; and subsequently these phases may impact on the reaction negatively by creating unfavourable routes for the reaction to proceed. On the other hand, decreasing the amount of $TiCl_4$ will only reduce the formation of TiH_2.

Design of Experiment (DOE) Analysis

The parameters and results of the DOE runs for the synthesis of TiH_2 are shown in Table 1. The two responses selected were CHNS analysis that represented in degree of dehydriding, $X_H(\%)$ and weight gain $X_W(\%)$ calculations collected for every experiments. As is shown in Table 1, the operating parameters were temperature, Ni catalyst percentage and time of the reaction. The results showed that the percentage of highest weight gain change ($X_w\%$) of the sample produced at 500 °C for 3 h with Ni catalyst of 50% was 89.11 wt% and the percentage of degree of hydriding ($X_H\%$) obtained to be 42.94% (Run #4). Moreover, increasing in temperature was increased the weight gain. It is obvious that the lowest weight gain was 15.40% (Run #3) when the reaction temperature was 300 °C.

Model graph analysis using perturbation plots was also performed in order to interpret the experimental results based on factorial model. Figure 3a and b shows the perturbation plot for both responses which are weight gain difference, $X_w(\%)$ and degree of dehydriding, $X_H(\%)$. From the figures shown, it can be observed that temperature and time have the positive effects; while, the Ni catalyst has the negative impact on $X_w(\%)$ and $X_H(\%)$. This was corroborated by the positive gradient of the perturbation plot for the factor temperature and time as well as negative gradient for the amount of catalyst factor. From the slopes of all three factors, the temperature factor showed the highest positive gradient which indicated the most significant parameter. Figure 3a shows that the temperature plays an important role in $X_w(\%)$. After temperature, the reaction time and the amount of catalyst influence the $X_w(\%)$. For the $X_H(\%)$ as it is shown in Fig. 3b, the temperature has the same effect. On the contrast, the time shows relatively less influence on the reaction. However, the factor of catalyst percentage played a significant role in affecting the $X_H(\%)$, whereas exhibited less effect on the weight gain.

The quadratic regression model for the two responses in terms of coded factors is demonstrated in Eqs. 13 and 14 and they were fitting well with the independent variables studied. These equations in terms of coded factors can be used to make predictions about the responses for the given levels of each factor. The coded equa-

Table 1 Percent Weight difference and Degree of Dehydriding for DOE Experimental Conditions

Run	A	B	C	$X_W(\%)$	$X_H(\%)$	Identified phases
1	300	10	3	21.62	32.59	49.3% CaClOH, 40.7% $Ca(OH)_2$, 10% Ni
2	500	10	3	66.23	41.87	16.4% CaClOH, 79.9% $Ni(TiO_3)$, 3.7% Ni
3	300	50	3	15.40	26.49	70% CaClOH, 28% Ni, 2% TiO_2
4	500	50	3	89.11	42.94	55.5% $CaCl_2(H_2O)_2$, 28.2% TiOCl, 16.3% Ni
5	300	10	5	25.71	34.10	–
6	500	10	5	55.02	40.64	–
7	300	50	5	16.44	29.06	47.9% CaClOH, 9.5% TiO_2, 16.3% $Ca(OH)_2$, 21.8% Ni, 4.4% TiO_2
8	500	50	5	77.59	42.70	82.1% $CaCl_2(H_2O)_4$, 8.6% TiOCl, 9.3% Ni
9	400	30	4	43.50	36.66	73.1% $CaCl_2(H_2O)_2$, 13.1% Ni, 0.3% $Ni_{3.68}Ti_{0.32}$
10	400	30	4	61.75	41.55	–
11	400	30	4	50.24	36.94	56.6% $CaCl_2(H_2O)_4$, 26.4% TiOCl, 3.9% $Ca(OH)_2$, 13.1% Ni

A: Temperature (°C). B: Ni Catalyst (%) C: Time (hours)
Weight gain: $X_w(\%)$. Degree of dehydriding: $X_H(\%)$

tions are useful for understanding the relative impact of the terms by comparing the terms coefficient in the equations.

$$\text{Degree of dehydriding, } X_H(\%) = 36.30 + 5.70A - 1.0B + 0.32C + 1.78AB - 0.69AC \quad (13)$$

$$\text{Weight gain, } X_w(\%) = 45.89 + 26.10A + 3.74B - 2.20C + 7.62AB - 3.48A \quad (14)$$

Where temperature is A, catalyst (%) is B, and time of reaction is C. The statistical significance of Eqs. 13 and 14 was also checked by F-test and the analysis of variance (ANOVA). The results showed that the model was highly significant and it gave a very low value of probability ($P < 0.05$) and the correlation coefficient (R^2) of 0.95 for weight gain and 0.97 for degree of dehydriding.

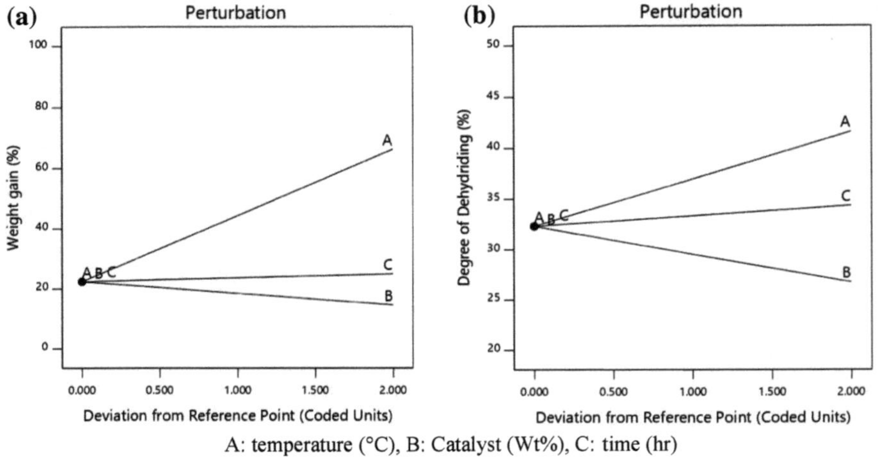

Fig. 3 **a** Most effective factor for weight gain, $X_W(\%)$ response and **b** degree of dehydriding, $X_H(\%)$

Microstructural Characterization

The reaction products were characterized to identify and confirm the expected phases formed during the dehydriding process. Of all the reactions listed in Table 1 as per the factorial design, experiments designated standard run #4 showed the highest $X_w(\%)$, run #3 showed the lowest $X_w(\%)$ and run #11 showed an average $X_w(\%)$. For these samples, the morphology, elemental analysis and the X-ray diffraction pattern were studied. For the sake of comparison, the reference sample was chosen with 30% catalyst. Figure 4 shows the EDS analysis of the mentioned samples. It can be seen that with decreasing weight gain the amount of Ti in the samples also decreased. This ascertains that the samples with higher $X_w(\%)$ had successfully trapped more Ti from $TiCl_4$ in the reduction process. By comparing the experimental parameters for these samples, it is obvious that the higher temperature resulted in the high $X_w(\%)$. In addition, all the samples had significant amount of oxygen atoms (O_2) as shown in EDS results. The reason could be related to the surface oxidation of the samples after each experiment.

The XRD analysis results are shown in Figs. 5, 6 and 7. Figure 5 shows two samples with the same catalyst percentage at 10% and the same reaction time for 3 h with different temperature. As observed, the peaks of CaH_2 have been all converted into other phases. Four phases have been detected in this setup containing Ni, CaClOH, $Ca(OH)_2$, $Ni(TiO_3)$. The oxidized phases, CaClOH, and $Ca(OH)_2$, indicate the presence of moisture and oxygen during or after the experiment. The moisture and oxygen may reach the samples during the reaction, after it was kept outside, or even during the preparation of XRD specimen. Visual investigation during the experiments shows no leakage of oxygen in the system. At 300 °C, $Ca(OH)_2$ phase

(a)

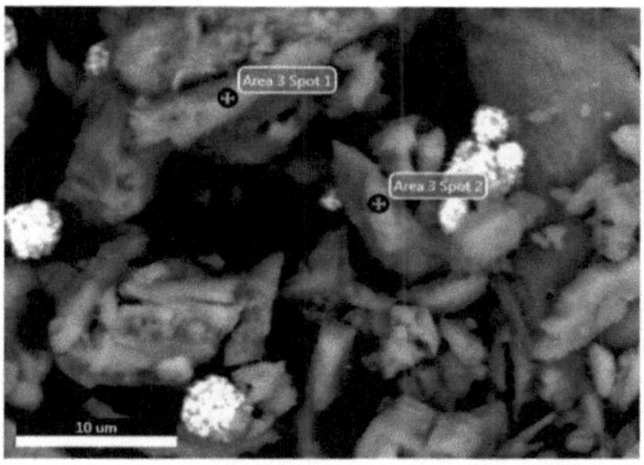

Area spot 1

Element	Weight %	Atomic %
C K	4.51	7.88
O K	52.97	68.32
Cl K	2.76	0.39
Ca K	40.57	21.25

Area spot 2

Element	Weight %	Atomic %
C K	5.56	8.31
O K	50.83	65.42
Ca K	43.62	22.78

(b)

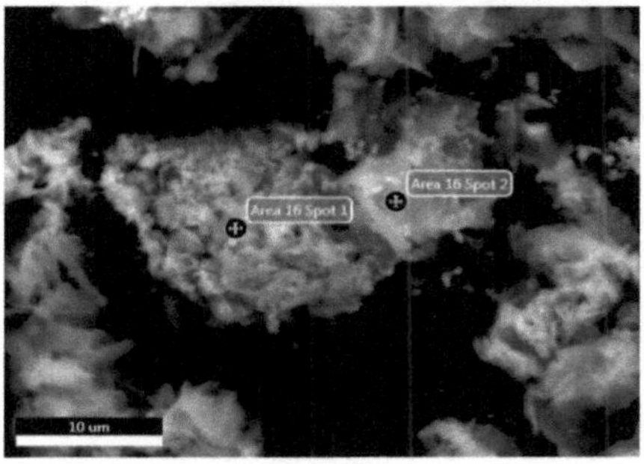

Area spot 1

Element	Weight %	Atomic %
C K	10.43	22.29
O K	20.49	32.89
Cl K	36.21	26.23
Ca K	11.19	7.17
Ti K	19.52	10.47
Ni K	2.16	0.95

Area spot 2

Element	Weight %	Atomic %
O K	16.13	33.60
Cl K	34.20	32.15
Ca K	6.42	5.34
Ti K	33.94	23.62
Ni K	9.31	5.28

(c)

Area spot 1

Element	Weight %	Atomic %
C K	24.27	43.27
Cl K	33.92	27.30
Ca K	38.92	27.21
Ti K	2.89	1.72

Area spot 2

Element	Weight %	Atomic %
C K	28.40	49.20
Cl K	36.45	28.51
Ca K	17.22	11.91
Ti K	17.94	10.38

(d)

Area spot 1

Element	Weight %	Atomic %
C K	4.54	9.60
O K	28.84	45.79
Cl K	30.76	22.04
Ca K	34.30	21.74
Ti K	1.56	0.82

Area spot 2

Element	Weight %	Atomic %
O K	32.44	53.26
Cl K	30.96	22.94
Ca K	34.98	22.92
Ti K	1.61	0.88

Fig. 4 EDS analysis, **a** Reference sample 62.5 wt% CaH_2, 18.75 wt% Ni, 18.75 wt% NH_4Cl, **b** sample run #4 with highest weight gain: 89.11%, **c** sample run #11 with average weight gain: 50.24%, **d** sample run #3 with lowest weight gain: 15.40%

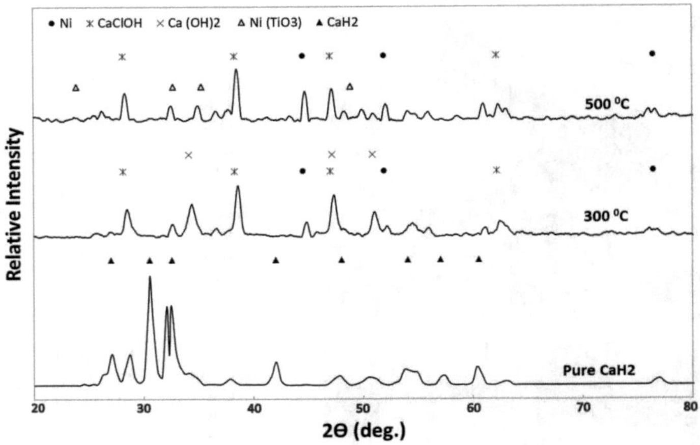

Fig. 5 XRD patterns comparison between Run #1 (300 °C) and Run #2 (500 °C) with pure CaH_2. Catalyst percentage and time were fixed at 10% and 3 h, respectively

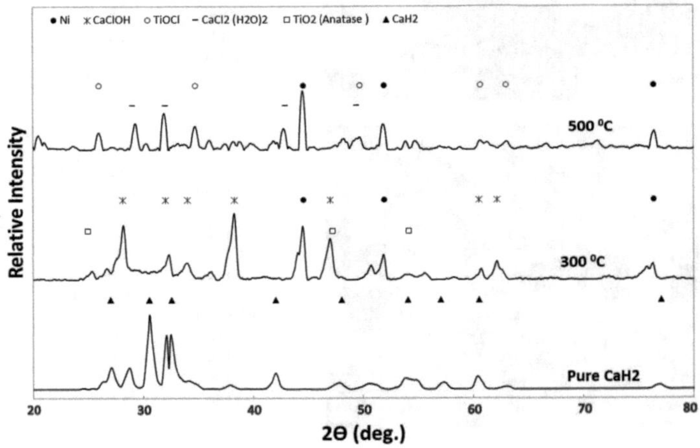

Fig. 6 XRD patterns comparison between Run #3 (300 °C) and Run #4 (500 °C) with pure CaH_2. Catalyst percentage and time were fixed at 50% and 3 h, respectively

detected which may result from the reaction between CaH_2 with moisture. In 500 °C, all the CaH_2 reacted to form TiH_2. However, this TiH_2 react with moisture creating TiO_2 and then with nickel to form $Ni(TiO_3)$ as Eq. 12 or directly with TiH_2 as Eq. 11. Both samples have CaClOH. This phase can be formed from $Ca(OH)_2$ if there was any $TiCl_4$ left over on the sample. Because $TiCl_4$ will react with moisture to create HCl, this HCl can react with $Ca(OH)_2$ and form CaClOH [17, 18]. The existence of Ti in the sample can increase the $X_w(\%)$ since the Ti has heavier molecular weight compared to the other elements in the system like Ca or Cl.

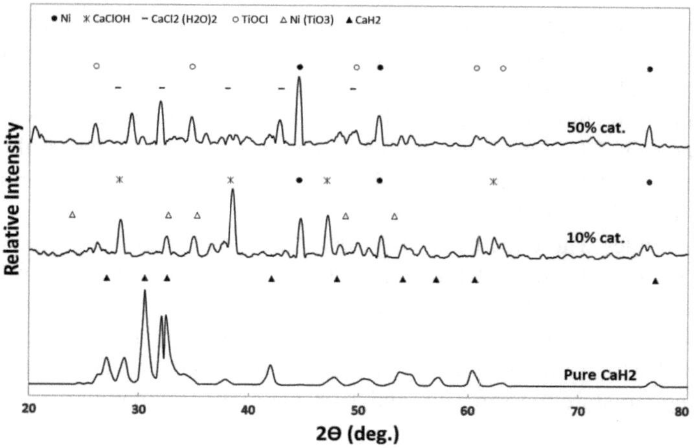

Fig. 7 XRD patterns comparison between Run #2 (10% catalyst) and Run #4 (50% catalyst) with pure CaH_2. Temperature and time were fixed at 500 °C and 3 h, respectively

In Fig. 6, the XRD analysis for run #3 and run #4 has compared. While a high amount of catalyst was used (about 50%) in these samples, the Ni is detected as major phase after experiments. A new phase was detected in these samples that can be associated with TiO_2 (anatase). The presence of TiO_2 could be related to the reaction between $TiCl_4$ and TiH_2 with moisture. As indicated from Fig. 6, there is no sign of $CaCl_2$ phase which means there was a little possibility of TiH_2 formation. The anatase phase is an unstable phase of TiO_2; however, this phase can be formed at the temperature below 300 °C [19]. In 500 °C for the run #4, $CaCl_2$ phase was detected. This could be related to the formation of TiH_2 as mentioned before based on the main equation of this study, Eq. 1. However, this phase has absorbed water to form $CaCl_2(H_2O)_2$. This formation is highly possible thermodynamically as shown in Eq. 15:

$$CaCl_2 + 4H_2O(g) = CaCl_2 \cdot 4(H_2O) \quad \Delta G^\circ = -80\,\text{kJ/mol at T} = 25\,^\circ\text{C} \quad (15)$$

Another phase which is detected in Fig. 6 at 500 °C is TiOCl. This phase can be formed only in temperature above 400 °C while reaction between $TiCl_4$ and $TiCl_3$ with moisture and oxygen occurs [20].

To determine the effect of catalyst in the same experiment conditions, the XRD analysis for run #2 and run #4 has compared in Fig. 7. From Fig. 7, the CaClOH phase can be seen only in 10% catalyst. As explained before, CaClOH needs $Ca(OH)_2$ that can be formed from CaH_2. In the sample with 50% catalyst, there was no CaClOH. The reason could be that the amount of Ni was too high at 50% catalyst which caused all the CaH_2 to be dehydrided and further prevented to form $Ca(OH)_2$ to help the formation of CaClOH. A study to a similar system shows dehydriding of MgH_2 with 5%wt. mol of Nickel [21]. On the other hand, a low amount of Ni has saved some

CaH_2 to be oxidized after the reaction and helping to form CaClOH. Moreover, the formation of $NiTiO_3$ is detected only in 10% catalyst. The low amount of Ni resulted in unreacted CaH_2 which could react as Eq. 1 to form TiH_2. However, $NiTiO_3$ can be formed only if the moisture exists with TiH_2 at low temperature as shown in Eq. 12. The existence of TiOCl in sample with 50% catalyst (Run #4) indicates the presence of moisture at high temperature [20]. Moreover, the $CaCl_2$ phase after formation was reacted with moisture to form $CaCl_2(H_2O)_2$.

In the samples where the catalyst is 10%, $CaCl_2$ and $NiTiO_3$ have been formed. This can be an indication of forming TiH_2 from the reaction between $TiCl_4$ and CaH_2. It should be mentioned that the authors previously proposed a new method to produce $TiCl_4$ through carbothermal reduction from ilmenite ore [22–24]. In the samples with a high amount of catalyst, CaH_2 is consumed, and $TiCl_4$ started to react with exiting moisture to form TiO_2 and TiOCl at higher temperature.

Conclusion

Experiments have been conducted to synthesis TiH_2 from the reaction between CaH_2 and $TiCl_4$ in the presence of Ni catalyst. The investigated parameters were temperature, amount of Ni catalysts and reaction time. The responses were weight gain, $X_w(\%)$ and the degree of dehydriding, $X_H(\%)$. The XRD analysis showed there was no detection of TiH_2. However, the other phases appearing in the diffractogram can be explained as precursor for the formation of TiH_2, i.e., appearance of by-product phases like $CaCl_2$. Moreover, TiH_2 was assumed to form $NiTiO_3$ by reacting with Ni and moisture in the humid atmosphere. This phase can be formed at room temperature. Also, TiO_2, in its both phases, anatase and rutile, can be formed above 300 °C between TiH_2 and moisture, and these two phases have been detected in XRD analysis. EDS spot analysis showed some titanium peaks in its spectrum which could be an indication for the existence of small amount of TiH_2 in the samples. However, the peaks were too small. The high amount of Nickel can be the reason behind this problem.

Design-Expert software was also used to analyse the interaction of the investigated factors and give good forecasting about $X_w(\%)$ and $X_H(\%)$ at different temperature, catalyst percentage, or reaction time change. The most effective parameter was observed is the temperature in changing the $X_w(\%)$, and catalyst percentage in change $X_H(\%)$. The less effective parameter in changing $X_w(\%)$ is catalyst percentage and the reaction time was the less effective parameter in changing the $X_H(\%)$.

Acknowledgements The authors wish to thank Universiti Sains Malaysia (USM) and Ministry of Higher Education (MOHE) of Malaysia for supporting this work. This research was supported primarily by the following grants; MOHE Fundamental Research Grant Scheme (FRGS) grant no: 203/PBAHAN/6071364 and USM Research University Individual (RUI) grant no: 1001/PBAHAN/814273. Further support by USM research grant was entitled Geran Bridging # 304/PBAHAN/6316116 and Nippon Sheet Glass Research Grant (NSGRG) (No. 304/PBA-

HAN/650360/N120). Special thanks to USM technician Mr. Encik Mohd. Helmi Bin Khir for supporting the author's experimental work.

References

1. Ezugwu EO, Bonney J, Yamane Y (2003) An overview of the machinability of aeroengine alloys. J Mater Process Technol 134(2):233–253
2. Boyer RR (2010) Attributes, characteristics, and applications of titanium and its alloys. JOM 62(5):21–24
3. Tal-Gutelmacher Ervin, Eliezer Dan (2004) Hydrogen-assisted degradation of titanium based alloys. Mater Trans 45(5):1594–1600
4. Olsson PAT, Blomqvist J, Bjerkén C, Massih AR (2015) Ab initio thermodynamics investigation of titanium hydrides. Comput Mater Sci 97:263–275
5. Gerdemann Steven J (2001) Titanium Process Technologies. Adv Mater Processes 159(7):41–43
6. Elias CN, Meyers MA, Valiev RZ, Monteiro SN (2013) Ultrafine grained titanium for biomedical applications: An overview of performance. J Mater Res Technol 3(4):340–350
7. Lütjering G, Williams JC (2007) Titanium, 2nd edn. Springer
8. Saboori A, Gallo D, Biamino S, Fino P, Lombardi M (2017) An overview of additive manufacturing of titanium components by directed energy deposition: microstructure and mechanical properties. J Appl Sci 7:883–906
9. Boyer RR, Williams JC, Wu X, Clark LP (2015) Titanium powder metallurgy: science, technology and applications: a realistic approach for qualification of PM applications in the aerospace industry. Elsevier, Pages 497–514
10. Duz V, Matviychuk M, Klevtsov A, Moxson V (2017) Industrial application of titanium hydride powder. Met Powder Rep 72(1):30–38
11. Udayakumar S, Sadaqi A, Ibrahim N, Fauzi MNA, Ramakrishnan S, Rezan SA (2018) Mathematical modelling of titanium hydride formation from titanium tetrachloride with magnesium hydride using matlab. J Phys: Conf Ser 1082:012037. http://dx.doi.org/10.1088/1742-6596/1082/1/012037
12. Udayakumar S, Sadaqi A, Ibrahim N, Fauzi MNA, Ramakrishnan S, Rezan SA (2018) Formation of titanium hydride from the reaction between magnesium hydride and titanium tetrachloride. J Phys: Conf Ser 1082:012003. http://dx.doi.org/10.1088/1742-6596/1082/1/012003
13. Azevedo CRF, Rodrigues D, Beneduce Neto F (2003) Ti–Al–V powder metallurgy (PM) via the hydrogenation–dehydrogenation (HDH) process. J Alloy Compd 353:217–227
14. Robertson IM, Schaffer GB (2010) Comparison of sintering of titanium and titanium hydride powders. Powder Metall 53(1):12–19
15. Weng Q, LI R, Yuan T, Shi Y, Qiu Z, Jiang M, He Y (2016) Hydrogenation reaction of metallic titanium prepared by molten salt electrolysis. Trans Nonferrous Met Soc China 26:1425–1432
16. Roine A (2006) Outokumpu HSC Chemistry, Chemical reaction and equilibrium software with extensive thermochemical database. Finland Outokumpu Research Oy
17. Allal KM, Dolignier JC, Martin G (1997) Determination of thermodynamical data of calcium hydroxichloride. Revue De L'institut Français Du Pétrole 52(3):361–368
18. Chen W-S, Chang F-C, Shen Y-H, Tsai M-S, Ko C-H (2012) Removal of chloride from MSWI fly ash. J Hazard Mater 237–238:116–120
19. Yin H, Wada Y, Kitamura T, Kambe S, Murasawa S, Mori H, Sakata T, Yanagid S (2001) Hydrothermal synthesis of nanosized anatase and rutile TiO_2 using amorphous phase TiO_2. J Mater Chem 11:1694–1703
20. Sekimoto H, Nose Y, Uda T, Sugimura H (2011) Preparation and properties of trivalent titanium compounds, $TiCl_3$ and TiOCl. High Temp Mater Processes (London) 30:435–440

21. Varin RA, Czujko T, Wronski ZS (2009) Thermal stability of Vale Inco nanonometric nickel asa catalytic additive for magnesium hydride (MgH_2). Int J Hydrogen Energy 34:8603–8610
22. Ahmadi E, Rezan SA, Baharun N, Ramakrishnan S, Fauzi A, Zhang G (2017) Chlorination kinetics of titanium nitride for production of titanium tetrachloride from nitrided ilmenite. Metall Mater Trans B 48(2017):2354–2366. https://doi.org/10.1007/s11663-017-1011-z
23. Ahmadi E, Shoparwe NI, Ibrahim N, Hamid SA, Baharun N, Ariffin KS, Hussin H, Fauzi MA (2018) The effects of experimental variables on iron removal from nitrided malaysian ilmenite by becher process, in, Springer International Publishing, Cham, pp. 1383–1396. http://dx.doi.org/10.1007/978-3-319-95022-8_113
24. Ahmadi E, Hamid SARBSA, Hussin HB, Baharun SRNB, Ariffin KSB, Fauzi MNA (2016) The preparation of iron-free $TiCl_4$ from reduced and nitrided ilmenite by polyethylene terephthalate. INROADS Int J Jaipur Natl Univ 5:11–16. http://dx.doi.org/10.5958/2277-4912.2016.00003.5

Nepheline Syenite—An Alternative Source for Potassium and Aluminium

Jayashree Samantray, Amit Anand, Barsha Dash, Malay K. Ghosh and Ajay K. Behera

Abstract An integrated approach has been made to extract potassium and aluminium from a silicate mineral, i.e. nepheline syenite with 5.4% K_2O, 19.9% Al_2O_3, 55.5% SiO_2. Chloridising-roasting experiments were carried using $CaCl_2$ at 900 °C for 1 h and water leaching at ambient temperature for 1 h. The leach liquor contains impurities like Ca and Na; hence, selective extraction of potassium was carried out through precipitation using perchloric acid at freezing temperatures. XRD data of the final KCl product obtained after decomposition of $KClO_4$ indicated cubical crystal system and was found to be 99.5% pure. In the second stage, leaching of the residue was carried out with HCl, H_2SO_4 and H_2SiF_6. For selective aluminium recovery, sulphuric acid was found to be the better leachant. A 2^3 full factorial design with time, temperature and acid concentration was used. Finally, aluminium sulphate crystals were produced from the leach liquor after removing iron through pH adjustment.

Keywords Nepheline syenite · Roasting · Leaching · Factorial design Aluminium sulphate

---

J. Samantray · B. Dash · M. K. Ghosh (✉)
Hydro and Electrometallurgy, CSIR-Institute of Minerals and Materials Technology, Bhubaneswar 751013, India
e-mail: mkghosh@immt.res.in

J. Samantray
e-mail: rubee.rai@gmail.com

B. Dash
e-mail: barshadash@immt.res.in

A. Anand
School of Minerals, Metallurgical and Materials Engineering, Indian Institute of Technology Bhubaneswar, Arugul 752050, India
e-mail: aa11@iitbbs.ac.in

J. Samantray · A. K. Behera
CSST, School of Chemistry, Sambalpur University, Jyoti Vihar, Sambalpur 768019, India
e-mail: ajaykumar.behera@yahoo.com

G. Azimi et al. (eds.), *Rare Metal Technology 2019*, The Minerals, Metals & Materials Series, https://doi.org/10.1007/978-3-030-05740-4_15

Introduction

Potassium is one of the three essential macronutrients for plants healthy growth along with nitrogen and phosphorous. It is involved in more than 60 enzymatic systems and is vital for the synthesis of proteins, vitamins, starch and cellulose. The distribution of potash in earth crust is not even. Around 90% of the potash minerals are distributed in Canada, Russia, Ukraine, Germany, Israel and America in the form of soluble K-chlorides or sulphates such as the minerals sylvite, sylvinite, kainite, and carnallite [1–3]. These minerals constitute K_2O content up to 65–75%. Extraction of potash from low-grade minerals such as feldspar, glauconitic sandstone, biotite and nepheline syenite containing 5–15% K_2O may be a viable option in countries like India [4–6].

Extraction of potassium from minerals may follow different routes. Acid leaching of glauconitic sandstone was carried out where it is reported that the initial stage of leaching was chemical controlled whereas the later stage of the reaction was diffusion controlled [7]. The chloridising-roasting of K-feldspar with calcium chloride was performed, and a systematic study was done by using various characterization techniques focusing on the effects of the calcination temperature and time, the particle size of the feldspar mineral, and the mass ratio of $CaCl_2$/K-feldspar mineral for the potassium extraction [8]. Melting of potash feldspar with salt additives is another innovative method for potassium extraction [9]. A combined reduction roasting-leaching method was developed by Shekhar et al. [10] to recover potassium values from glauconitic sandstone suitable for fertilizer application.

Selective separation of potassium through precipitation is really tricky because potassium in almost all forms is soluble in water. Marcus and Asher [11] reported the extraction of potassium halides from their aqueous solutions by dibenzo-18-crown-6 and dicyclohexo-18-crown-6 ethers dissolved in protic solvents, which solvate the anions effectively. Moreover, crown ethers are water miscible. It requires another water immiscible solvent in which crown ether is soluble and thus enables it to extract alkali metals. Burungale [12] developed a method for solvent extraction of potassium from 0.001 M picric and 0.001 M Dicyclohexano-18-crown-6 in nitrobenzene. Marcus [13] tried to extract potassium from brine through dibenzo-18-crown-6 with substituted phenol as diluents.

George et al. patented a process for the separation of potassium from potassium-bearing brine by contacting it with $NaClO_4$ to precipitate out $KClO_4$ which was then separated and formed into solution. The potassium-rich solution was separated through ion exchange method with sodium, after which the resultant free potassium is combined with chloride, sulphate, nitrate or carbonate ion [14].

The mechanism of aqueous alteration of alumina-silicate minerals and the chemical processes for the extraction of potassium and aluminium were also reported [15]. Production of alumina from nepheline ore in Razgah deposit was done by studying chemical composition, size reduction, sintering, leaching and de-silication processes, carbonization and crystallization of aluminium hydroxide [16]. Complex processing of non-bauxite ores for obtainment of aluminium and alumina with the production of amorphous silica, silicon, and other useful components was developed by Rimkevich et al. [17].

A complete process of extraction and purification of the potassium salt from the nepheline syenite has been studied. First part deals with the leaching of potassium from the sample through chloridising-roasting followed by water leaching method. Then, potassium was selectively extracted from leach liquor through precipitation method using perchloric acid at freezing temperatures. Potassium precipitated as potassium perchlorate. Finally, crystals of potassium chloride were prepared by roasting $KClO_4$ in the muffle furnace for 1 h at 600 °C.

Acid leaching of residue was done for the extraction of aluminium using HCl, H_2SO_4, and H_2SiF_6. Among the three acids, H_2SO_4 leach liquor was found better for aluminium recovery. Three parameters were chosen for a 2^3 full factorial design which were time, temperature and acid concentration. As the liquor constitutes iron along with aluminium, Fe was removed by pH adjustment to get aluminium-rich liquor.

Experimental

Materials

The nepheline syenite (NS) sample used in this study was collected from Koraput district of Odisha, India. The ore was crushed, ground and sieved to B.S.S. 170 mesh. Calcium chloride dihydrate and perchloric acid were obtained from Merck (India) and HCl, H_2SO_4 and ethylene glycol from HiMedia Laboratories (India). Double distilled water was utilized for making solutions and dilutions. Fluorosilicic acid used in the experiments was obtained from a fertilizer industry of Odisha, India.

Methods

Roasting and Leaching

Di-hydrate $CaCl_2$ was mixed uniformly with ground nepheline syenite sample in the ratio NS:$CaCl_2$ of 1:0.45 and was transferred to alumina crucibles. The crucibles containing the mixture were subjected to roasting in a muffle furnace at 900 °C for 1 h. The roasted product was naturally cooled to room temperature inside the furnace. The product was then removed from the crucible, ground in a mortar and pestle. The ground samples were leached at room temperature in a glass beaker kept over a magnetic stirrer for 30 min, and stirring speed was maintained at 300 rpm. Then, the leached slurry was filtered, and the filtrate was used for potassium analysis in the flame photometer.

Precipitation

The leach liquor containing K was taken in a double-walled glass reactor connected to Julabo® thermostatic water bath. Total temperature range studied during the process was −7 to 7 °C. To attain the temperature between 1 and 7 °C, water was taken in the Julabo water bath and maintained in the preset temperatures. When the liquor attained the set temperature, perchloric acid was added. To obtain the freezing temperatures from 0 to −7 °C azeotropic mixture of ethylene glycol (60%) and water (40%) was circulated in the glass reactor in place of water.

The precipitate was filtered, thoroughly washed and oven-dried at 80 °C overnight and then characterized using XRD technique.

Residue Leaching and Iron Removal

The residue obtained after water leaching was subjected to leaching with three acids such as HCl, H_2SO_4 and H_2SiF_6. Experiments with HCl and H_2SO_4 were conducted in double-walled glass reactor whereas experiments with fluorosilicic acid were carried out in Teflon reactor set-up with Teflon stirring rod and impeller. The Teflon reactor set-up was kept inside a static thermostatic water bath with preset temperature. A 2^3 full factorial design was performed for sulphuric acid leaching to know the effects of various parameters and obtain maximum aluminium content during leaching.

The leach liquor obtained by sulphuric acid leaching contained Fe along with Al which needed purification step for the removal of Fe. This process was carried out at room temperature having initial pH of the solution as 1.1 from which pH was raised to 3.5 using 40 grams of $Ca(OH)_2$ for 1L of leach liquor.

Analytical Method

Wet chemical methods were used for the elemental analysis of the ore such as Si, Al and Fe. Contents of alkali metals K and Na were determined using flame photometer (Elico India-Model CL378); Ca content was determined by ICP-OES (Perkin Elmer-Model Optima 8300) and other metals in Atomic Absorption Spectroscopy (Perkin Elmer-Model 2380). The mineral phases of the sample at various experimental stages were detected in X-Ray diffractometer (PANalytical-Model X'Pert PRO).

Results and Discussion

Characterizations

Chemical composition of ground nepheline syenite powder has been presented in Table 1.

Nepheline syenite mineral is medium to coarse-grained holocrystalline, feldspathic, plutonic igneous rock largely made up of nepheline, sodium feldspar (albite), alkali feldspar (orthoclase, microcline) and hematite. The structure of nepheline is based on six-member rings of silica tetrahedra with apices alternating in direction, forming hexagonal and polygonal voids, and it belongs to the hexagonal crystal system [18]. The X-Ray diffraction (XRD) pattern of the nepheline syenite ore is given in Fig. 1.

The phases present are **microcline** ($KAlSi_3O_8$) [ICDD ref. no.: 00-019-0932], **nepheline** (Na, K) $AlSiO_4$ [ICDD ref.no.: 01-070-1582], **albite** ($NaAlSi_3O_8$) [ICDD ref.no.: 01-089-6429] and **hematite** (Fe_2O_3) [ICDD ref. no.: 01-073-0603].

Alumino-silicate framework is very stable and hard to break through simple leaching process using any acids or bases. It is believed that potassium occupies the space between two layers. Hence, release of potassium is only possible through chemical attack of the silicate matrix.

Table 1 Composition of nepheline syenite

Compound	K_2O	SiO_2	Al_2O_3	Na_2O	CaO	Fe_2O_3	LOI	Others
wt%	5.4	55.5	19.79	11.13	3.8	2.81	1.02	0.55

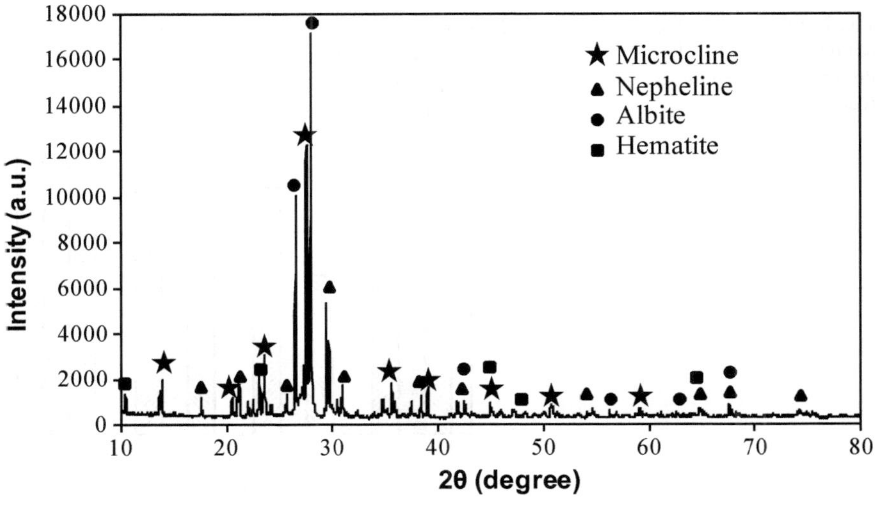

Fig. 1 XRD patterns of nepheline syenite powder sample

Roasting-Leaching Studies

Effect of Roasting Temperature and $CaCl_2$ Dosage

The effect of roasting temperature on the extraction of potassium was studied in the range 700–1000 °C, keeping the roasting time of 1 h and water leaching conditions constant. The results are presented in Fig. 2, which show significant effect of temperature on potassium extraction in the temperature range of 700–900 °C, with potassium extraction reaching 99.19% at 900 °C. Increasing the temperature beyond 900 °C had a little effect on extraction of potassium. It is also evident that ratio of NS: $CaCl_2$ has significant positive effect on the recovery of potassium. By increasing weight percentage of calcium chloride from 15 to 45%, potassium recovery increased from 57.8 to 99.19% at 900 °C. So nepheline syenite to $CaCl_2$ ratio of 1:0.45 and 900 °C are taken as the optimized roasting conditions for maximum potassium recovery.

After roasting, the major phases formed were sylvite (KCl), anorthite ($CaAl_2Si_2O_8$) and wollastonite ($CaSiO_3$). As the $CaCl_2$ melts at 860 °C, all the chloridising-roasting experiments were carried at 900 °C. The possible reaction for synthesis of various compounds from chloride roasting of NS is as follows:

$$\mathrm{Na_3KAl_4Si_4O_{16}} + 0.5O_2 + 3CaCl_2 \rightarrow CaAl_2Si_2O_8 + 2CaSiO_3 + 3NaCl + KCl + Al_2O_3 + Cl_2 \quad (1)$$

After water leaching, the generated KCl dissolved forming the potassium-rich leach liquor leaving behind the residue containing anorthite ($CaAl_2Si_2O_8$) and wollastonite ($CaSiO_3$) and Al_2O_3 which are water insoluble.

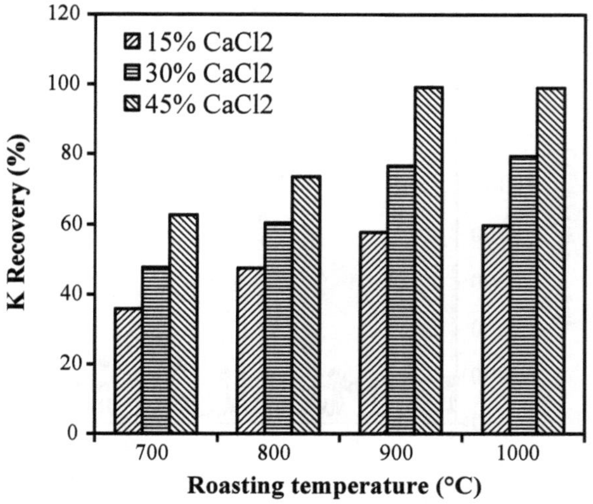

Fig. 2 Effect of roasting temperature and $CaCl_2$ dosage on potassium recovery

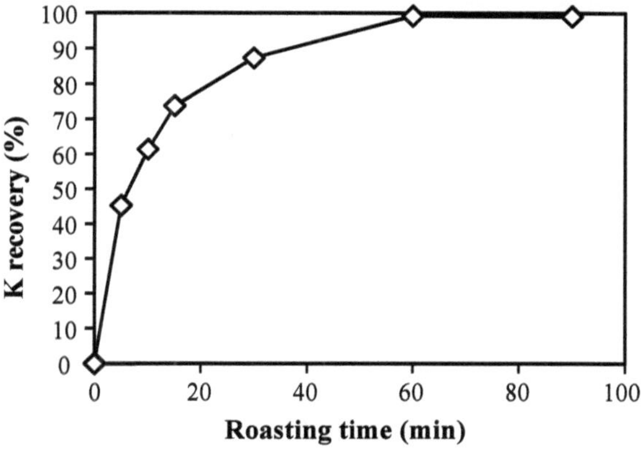

Fig. 3 Effect of roasting time on potassium extraction

Effect of Roasting Time

In order to study the role of roasting time on the extraction of potassium, the time was varied from 5 to 90 min while keeping the temperature at 900 °C, ratio of NS: $CaCl_2$ as 1:0.45 and water leaching condition constant. The results shown in Fig. 3 indicate that a roasting time of 1 h is sufficient to release maximum amount of potassium.

Selective Precipitation of $KClO_4$

The leach liquor after first-stage leaching of chloridized roast product contained significant amount of calcium and sodium along with potassium. Almost all salts of potassium are water soluble; selective extraction of potassium can only be feasible by precipitating potassium as perchlorate salt using stoichiometric amount of perchloric acid. The precipitation process was effective at freezing temperatures. From Fig. 4, we observe that the lower the temperature is, the higher the percentage precipitation is. At −7 °C, we achieved 99.92% of potassium recovery as potassium perchlorate using stoichiometric amount of perchloric acid.

Production of KCl

Bulk use of potassium perchlorate for agricultural purpose is not feasible as it is nearly insoluble in water. Also, other uses of $KClO_4$ are rather limited. Hence, it is essential to convert the insoluble $KClO_4$ into some water-soluble potassium salts.

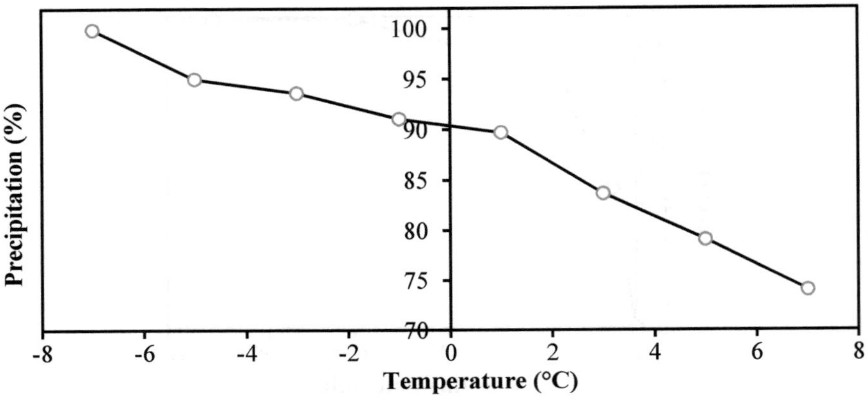

Fig. 4 Effect of temperature on precipitation of potassium

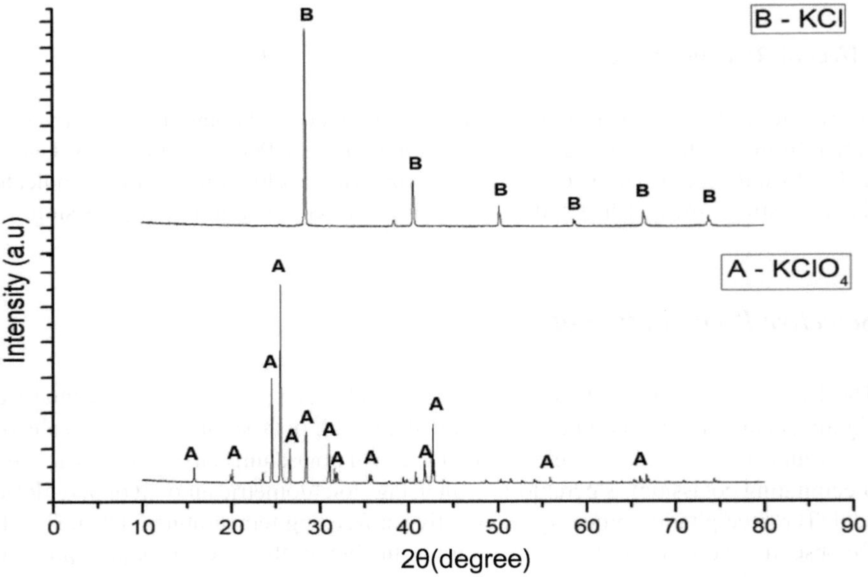

Fig. 5 XRD patterns of $KClO_4$ and KCl crystals

Hence, thermal decomposition of $KClO_4$ at its decomposition temperature of 600 °C was carried out in the muffle furnace for 1 h.

Decomposition reaction

$$KClO_4 \longrightarrow \Delta KCl + 2O_2 \tag{2}$$

White crystalline powder of KCl was obtained, and it was confirmed from the XRD that decomposition of $KClO_4$ gives KCl. The XRD patterns of the KCl (Fig. 5) are in very good agreement with standard KCl which possesses face-centred cubic crystal structure. The 100% peak intensity [002] refers to d-spacing of 3.14 Å at 2θ value of 28.31. The purity of the KCl is estimated as 99.5%.

Aluminium Recovery

Preliminary Leaching Tests

Three acids, viz. HCl, H_2SO_4 and H_2SiF_6, of strength 1 N were employed to screen the most suitable leaching agent for the recovery of aluminum from 1st-stage leach residue. Screening tests conducted at 90 °C, 10% pulp density for various time periods as shown in Table 2 demonstrate that both sulphuric acid and hydrochloric acid result in the same degree of aluminium recovery whereas fluorosilicic acid leaches less efficiently. But H_2SO_4 is more preferable and suitable leaching agent for aluminium in the present case because it generates a leach liquor with almost no calcium, which is beneficial since it will eliminate the additional calcium removal step during downstream processing.

Factorial Design of Experiments

Three main factors considered for the factorial design of experiments are temperature, time and concentration of H_2SO_4. The combinations at which experiments were conducted and corresponding recovery values are given in Table 3. It is observed that maximum extraction of 95.2% is obtained at [1] levels of the coded factors A, B and C, respectively, i.e., time, temperature and concentration of H_2SO_4, respectively.

Leaching extent achieved at design conditions was analysed statistically to generate ANOVA data and other plots. Following information can be inferred from analysis of variance table and the plots:

1. The main effect plot (Fig. 6) indicates that higher concentration of H_2SO_4 results in the higher aluminium extraction whereas time and temperature have moderate effect on aluminium recovery.
2. It is well known that if P is ≤ 0.05, then the term will be statistically significant. In Table 4, it is observed that concentration (C) is the most significant term followed by time (A) and temperature (B). All the interaction terms are significant except time * concentration (A * C).

Table 2 Leaching efficiency of acids for Al recovery

Time (min)	Leaching (%)								
	HCl (1 N)			H_2SiF_6 (1 N)			H_2SO_4(1 N)		
	Al	Fe	Ca	Al	Fe	Ca	Al	Fe	Ca
30	51.21	21.41	11.3	39.79	16.88	5.47	53.1	18.23	Nil
60	64.17	27.37	17.23	57.03	24.69	7.31	66.1	23.1	Nil
90	75.31	32.3	23.1	67.9	28.8	9.37	78.3	30.21	Nil
120	89.71	37.12	27.6	79.29	33.25	13.91	91.7	32.32	Nil

Table 3 Operating conditions for experiments

Coded factor for the parameters			Experimental conditions			Al recovery (%)
Time (A)	Temperature (B)	H_2SO_4 Conc. (C)	Time (h)	Temperature (°C)	H_2SO_4 Conc. (M)	
Levels						
−1	−1	−1	0.5	30	0.5	33.85
1	−1	−1	1.5	30	0.5	53.85
−1	1	−1	0.5	90	0.5	48.87
1	1	−1	1.5	90	0.5	49.92
−1	−1	1	0.5	30	1.5	71.15
1	−1	1	1.5	30	1.5	84.25
−1	1	1	0.5	90	1.5	83.60
1	1	1	1.5	90	1.5	95.20
0	0	0	1.0	60	1.0	66.10
0	0	0	1.0	60	1.0	64.30

Table 4 Analysis of variance

Source	DF	Adj SS	Adj MS	*F*-value	*P*-value
Model	7	3248.55	464.08	565.71	0.002
Linear	3	3137.61	1045.87	1274.90	0.001
A	1	261.63	261.63	318.93	0.003
B	1	148.70	148.70	181.26	0.005
C	1	2727.28	2727.28	3324.53	0.000
2-way interactions	3	72.88	24.29	29.61	0.033
A*B	1	52.28	52.28	63.72	0.015
A*C	1	1.67	1.67	2.03	0.290
B*C	1	18.94	18.94	23.09	0.041
3-way interactions	1	38.06	38.06	46.40	0.021
A*B*C	1	38.06	38.06	46.40	0.021
Error	2	1.64	0.82		
Curvature	1	0.02	0.02	0.01	0.928
Pure error	1	1.62	1.62		
Total	9	3250.19			

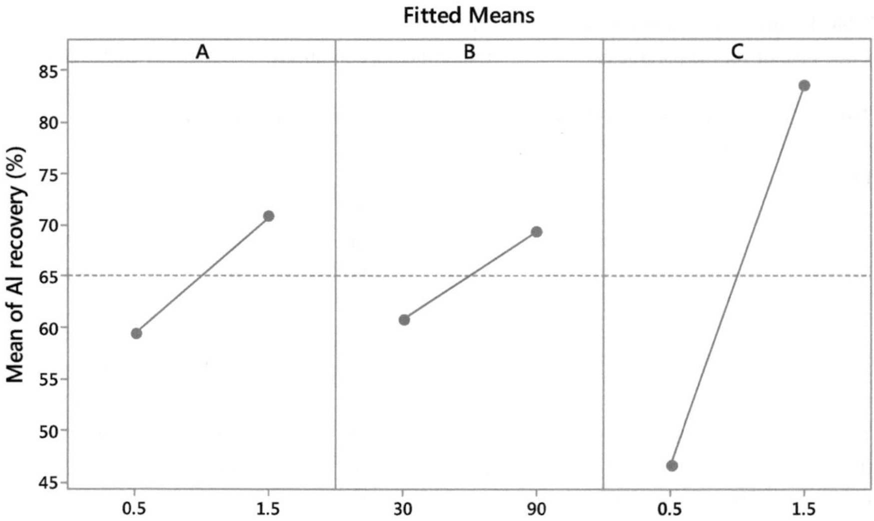

Fig. 6 Main effects plot for Al recovery (%) in H_2SO_4 leaching

Table 5 Validation of regression equations for Al

Time (h)	Temperature (°C)	Concentration (M)	Extent of leaching (%)	
			Experimental	Calculated
0.6	50	0.6	43.7	43.88
0.75	45	0.75	50.3	50.45
1.1	70	1.1	71.1	71.35
1.25	75	1.25	79.2	79.48

Regression Equation in Uncoded Units

$$\text{Al recovery (\%)} = -11.57 + 37.29\,A + 0.5024\,B + 46.40\,C - 0.4613\,A*B - 15.63\,A*C - 0.1883\,B*C + 0.2908\,A*B*C \quad (3)$$

Additionally, some random experiments were performed to validate the generated regression equation (Eq. 3). The experimental and the predicted values of leaching extent achieved at those random conditions were in close proximity (Table 5). Thus, it can be concluded that the generated equation is capable of predicting leaching extent within $\pm 2\%$ of the actual values.

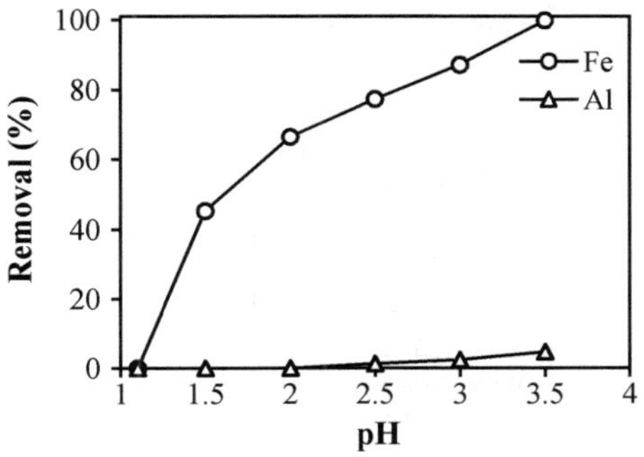

Fig. 7 Iron removal at various pH

Iron Removal

As the iron was present in its +3 oxidation state, it can be removed by increasing the pH up to 3.5. $Ca(OH)_2$ was chosen over other neutralizing agents as it increases the pH in the sulphate liquor without introducing additional metal ions into the system as it forms insoluble gypsum during the neutralization process. Initially when the pH was raised from 1.1 to 2, there was no effect on the aluminum co-precipitation while iron precipitation steadily increased. On reaching the pH up to 2.5, iron precipitation was more than 75%, whereas that of aluminum was less than 2%. Finally, on reaching pH 3.5, iron was completely removed from the leach liquor with aluminum loss of <5% presumably as co-precipitation (Fig. 7).

The Al-rich liquor can be further processed to get the Al_2SO_4 crystals through crystallisation or spray drying. A conceptual flow sheet is shown in Fig. 8 for effective recovery of K and Al from nepheline syenite.

Conclusion

Potassium is an essential micronutrient for plants. Besides high-grade resources like sylvite and sylvanite, it can also be obtained from low-grade potassium-bearing minerals, such as nepheline syenite. Using $CaCl_2$ as chloridising agent in the NS: $CaCl_2$ ratio of 1:0.45, about 99% potassium values could be recovered from nepheline syenite by roasting at 900 °C for 1 h. The K-content from the alumino-silicate structure was converted to soluble form enabling it to dissolve by simple water leaching. Selective extraction of K from leach liquor containing Ca, K, Na, etc. was feasible through precipitation using perchloric acid at freezing temperature of −7 °C using ethylene glycol-water mixture as heat transfer medium. The KCl crystal obtained through calcination of $KClO_4$ crystals at 600 °C was of 99.5% purity which could be confirmed by comparing with XRD patterns of standard KCl. The first-stage leach residue could

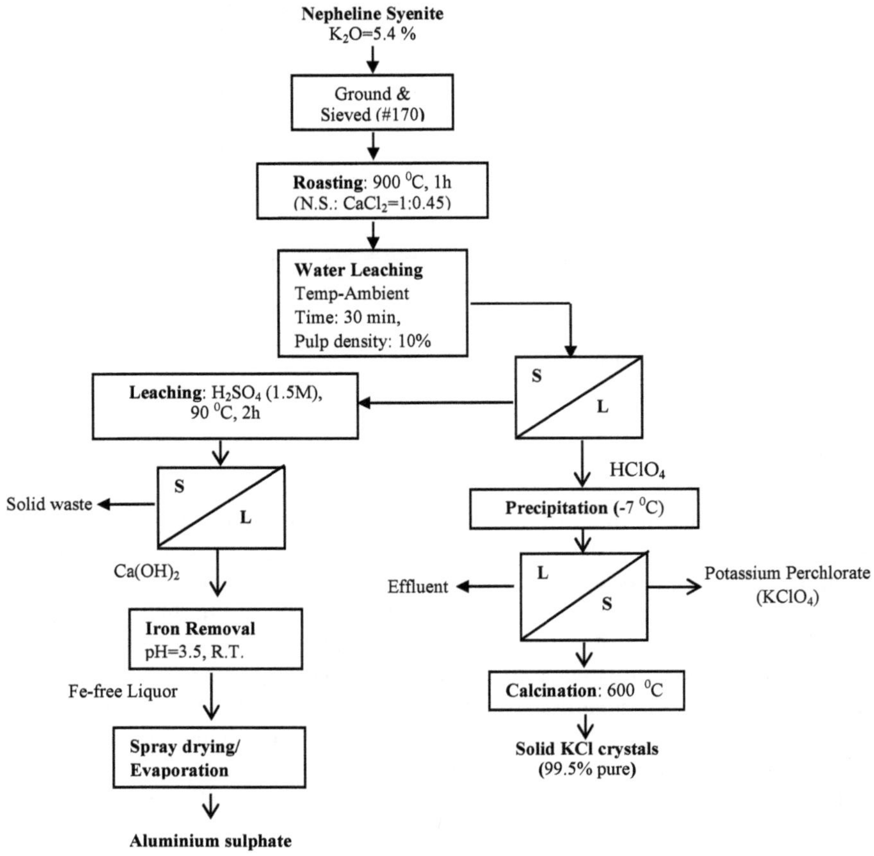

Fig. 8 Conceptual process flow sheet for recovery of potassium and aluminium

be used to recover aluminum values through H_2SO_4 leaching. Through a 2^3 full factorial design, effects of acid concentration, time and temperature demonstrated that most significant parameter is acid concentration. More than 95% aluminium could be extracted under the leaching conditions H_2SO_4 1.5 M, temperature 90 °C, time 1.5 h and pulp density 10% (w/v). Iron from the acid leach liquor could be removed effectively by increasing the pH to 3.5 using $Ca(OH)_2$. The obtained aluminium-rich liquor can be spray-dried to get $Al_2(SO_4)_3$ crystals.

Acknowledgements The authors are thankful to the Director, CSIR-IMMT Bhubaneswar for his kind permission to publish this work. One of the authors (JS) expresses her gratitude to CSIR for providing the fellowship under XII plan period projects. The authors also wish to gratefully acknowledge the assistance rendered by Mr. Pravat K. Behera for XRD studies.

References

1. Mansfield GR (1922) Potash in the greensands of New Jersey. USGS Report, Bulletin 727, Washington
2. Ståhlberg S (1959) Studies on the release of bases from minerals and soils. Acta Agric Scand 9:448–456. https://doi.org/10.1080/00015125909436218
3. Manning D, Limon MS, Brandt K (2014) Mineralogical controls on K fertiliser function. In: 16th World Fertiliser Congress of CIEC, pp 132–134
4. Yadav VP, Sharma T, Saxena VK (2000) Dissolution kinetics of potassium from glauconitic sandstone in acid lixiviant. Int J Miner Process 60:15–36
5. Varadachari C (1997) Potash fertilizer from biotite. Ind Eng Chem Res 36:4768–4773
6. Rao JR, Nayak R, Suryanarayana A (1998) Feldspar for potassium, fertilisers, catalysts and cement. Asian J Chem 10:690–706
7. Mazumder AK, Sharma T, Rao TC (1993) Extraction of potassium from glauconitic sandstone by the roast-leach method. Int J Miner Process 38:111–123
8. Yuan B, Chun L, Liang B, Lü L, Hairong Y, Haoyi S, Longpo Y, Heping X (2015) Extraction of potassium from K-feldspar via the $CaCl_2$ calcination route. Chin J Chem Eng 23:1557–1564. https://doi.org/10.1016/j.cjche.2015.06.012
9. Hao Z, Desi S, Hong B (2012) The extraction of potassium from feldspar by molten salt leaching method with composite additives. Adv Mater Res 524–527:1136–1139. https://doi.org/10.4028/www.scientific.net/AMR.524-527.1136
10. Shekhar S, Mishra D, Agrawal A, Sahu KK (2017) Physico-chemical treatment of glauconitic sandstone to recover potash and magnetite. J Clean Prod 147:681–693. https://doi.org/10.1016/j.jclepro.2017.01.127
11. Marcus Y, Asher LE (1978) Extraction of alkali halides from their aqueous solutions by crown ethers. J Phys Chem 82:1246–1254
12. Burungale SH (2014) Rapid solvent extraction of potassium (I) with dicyclohexano-18-crown-6 from picrate medium. Int J Curr Res 6:8412–8417
13. Marcus Y (1977) The activities of potassium chloride and of water in Dead Sea brine. Geochim Cosmochim Acta 41:1739–1744
14. George DAR, Riley JM, Ross JR (1969) Method for recovering and producing potassium salts. US Patent no. 3,429,657
15. Skorina T, Allanore A (2015) Aqueous alteration of potassium-bearing aluminosilicate minerals: from mechanism to processing. Green Chem 17:2123–2136. https://doi.org/10.1039/c4gc02084g
16. Jorjani E, Amirhosseini M (2007) Alumina production process from nepheline ore in Razgah (Iran). In: Mineral Processing Technology (MPT 2007), pp 111–115
17. Rimkevich VS, Malovitskii YN, Bogidaev SA, Pushkin AA, Dem'yanova LP, Eranskaya TY (2008) Effective technologies for complex processing of non-bauxite ores. Russ J Non-Ferrous Met 49:97–103. https://doi.org/10.3103/s1067821208020065
18. Mclemore VT (2003) Nepheline syenite. In: Kogel JE, Trivedi NC, Barker JM, Krukowski ST (eds) Industrial minerals, 7th edn. Society for Mining, Metallurgy, and Exploration, Littleton, pp 653–670

Part IV
Rare Metals IV

Recovery Vanadium from Vanadium-Bearing Hazardous Residues

Zhigan Deng, Xingbin Li, Chang Wei, Gang Fan, Cunxiong Li and Minting Li

Abstract The extracted vanadium residues of vanadium titanomagnetite is one of the most hazardous residues that have a large annual output. In the paper, vanadium was leached by composed leaching reagent of sulphuric acid, hydrofluoric acid and potassium permanganate from the residues. The process parameters such as reagent concentration, ratio of liquid to solid, temperature, leaching time, which impact on the vanadium leaching percent, were investigated. The results show that HF can destroy the silicate phase which surrounds the vanadium-iron spinel. The addition of $KMnO_4$ can enhance the oxidation transformation of insoluble vanadium (*trivalence*) to soluble vanadium (*tetravalence*) or vanadium (*pentavalence*), thus improve the vanadium leaching percent to 85%, under the reaction conditions: granule size of 0.15–0.25 mm, 125 g L^{-1} H_2SO_4, 30 g L^{-1} HF, 3 wt% $KMnO_4$, liquid-to-solid ratio of 5 mL g^{-1}, at 90 °C for 4 h.

Keywords Vanadium-bearing residue · Composed reagent · Leaching Oxidation transformation

Introduction

Vanadium is an essential rare metal. Vanadium is widely distributed and in the nature, but it is not present alone mineral deposit of vanadium, it generally occurs in combination with various minerals which include carnotite, vanadium-titanium magnetite, roscoelite, vanadinite, mottramite and patronite as important sources of the metal [1–3]. Most of the world's vanadium production originates from titanomagnetite ores. But millions of tons of extracted vanadium residue are produced at the same time every year [4–6]. Those residues contain a notable amount of heavy metals, and simple disposals of these wastes will cause some environmental problems and

Z. Deng (✉) · X. Li · C. Wei · G. Fan · C. Li · M. Li
Faculty of Metallurgical and Energy Engineering, Kunming University of Science and Technology, 253 Xuefu Road, Kunming 650093, Yunnan, China
e-mail: dengzhigan83@163.com

G. Azimi et al. (eds.), *Rare Metal Technology 2019*, The Minerals, Metals & Materials Series, https://doi.org/10.1007/978-3-030-05740-4_16

increases the disposal costs. As environmental concerns increase and legislation regarding hazardous residues forces companies to process their own products and residues, various attempts have been made to use the residues. Some research has indicated that the vanadium contained in the extracted residue is about 1.5 wt% [7–9], and the content of vanadium in the extracted residue is higher than that in titano-magnetite ore which contained vanadium is about 0.5 wt% [3]. The total amount of vanadium contained in the residue is considerable. Therefore, it is desirable to recover the valuable vanadium from these waste residues.

Some research [4–6, 10] has shown that the vanadium in the vanadium residue consisted of vanadium-iron spinel and silicates is present as insoluble V^{3+}. The vanadium-iron spinel and the silicates are the main mineral phases that contain vanadium in the vanadium residue and the vanadium-iron spinel was mostly enclosed by a silicate phase as well. The surrounding silicate phase hindered oxygen mass transfer, diffusion and the formation of vanadate. Therefore, to leach the vanadium contained within vanadium-iron spinel, the surrounding silicates need to be destroyed first and then the vanadium in the vanadium-iron spinel can be oxidized and leached.

Roasting is an effective method to destroy silicate phases [11, 12]. However, the addition of a sodium salt such as NaCl, Na_2CO_3 or $NaHCO_3$ is usually required, which produces poisonous gases such as HCl and Cl_2 during the roasting process resulting in serious environmental pollution. Apart from roasting, hydrofluoric acid can also effectively decompose the silicate phase [13, 14].

In this paper, authors used hydrofluoric acid to destroy the silicate and then leached vanadium with sulfuric acid and $KMnO_4$ to oxidize the insoluble V^{3+} to soluble V^{4+} and V^{5+}, with the purpose of leaching vanadium. The contributions in this work are to determine the optimum conditions for vanadium extraction, leaching behaviour of vanadium, and provide a technology of recovery vanadium of vanadium-bearing hazardous residues.

Materials and Methods

Materials

The experimental work has been performed on a sample come from the industrial extracted vanadium residue. It was collected from waste disposal sites at a vanadium plant of the Panzhihua iron and steel company in China. The extracted vanadium residue was dried to a constant weight at 60 °C, after crushed and grinded. The main components of the vanadium slag are listed in Table 1. And the XRD pattern of extracted vanadium residue is shown in Fig. 1. The XRD results show that hematite (Fe_2O_3), magnetite (Fe_3O_4) and vanadium-iron spinel (Fe_2VO_4) were the main mineral components of iron and vanadium. All the reagents used in our experiments were of analytical grade.

Table 1 Chemical compositions of extracted vanadium residue

Components	wt%	Components	wt%
V	1.293	Na_2O	5.06
Fe_{metal}	<0.5	MgO	5.78
Fe_{total}	28.53	Cr	1.42
Ti	7.19	CaO	2.34
SiO_2	16.42	S	0.042
Mn	6.30	K_2O	0.10
Al_2O_3	3.42	P	0.046

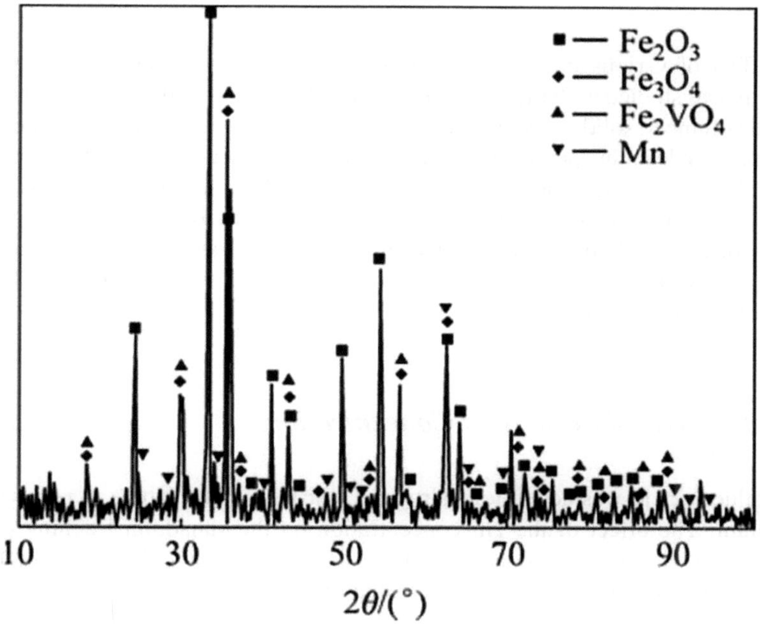

Fig. 1 XRD pattern of extracted vanadium residue

Methods

The method of this thesis is that leaching process was performed in agitated flasks with a reflux condenser using mechanical agitation at a speed of 500 rpm. The leaching temperature was controlled by water-bath heating. We added dilute sulfuric, hydrofluoric acid and potassium permanganate to the extracted vanadium residue at a pre-determined liquid/solid ratio. After the required contact time, the suspension was filtered and the filter cake was washed with water until the filtrate was colorless. Finally, the filter cake was dried at 60 °C for an analysis of its vanadium and other chemical content. The leaching rate η was calculated using the following equation:

$$\eta = \left(1 - m_2\omega_2 / m_1\omega_1\right) \times 100\% \quad (1)$$

Where m_1 and m_2 are the raw material and filter cake weights, respectively, ω_1 and ω_2 are the vanadium content of the raw material and the filter cake, respectively.

Results and Discussion

Effect of Sulfuric Acid Concentration

The effect of various initial sulfuric acid concentrations on percent of vanadium leaching was studied with the same condition of the 20 g L^{-1} HF, 1 wt% $KMnO_4$, L/S (liquid volume to solid mass ratio) of 5 mL g^{-1} at 90 °C for 4 h. Figure 2 shows that the vanadium leaching percent increased as the sulfuric acid concentration increased. At a sulfuric concentration of 87.5 g L^{-1}, the vanadium leaching ratio was 58.2%, while at a sulfuric acid concentration of 125 g L^{-1} it was 70.5%. A further increase in the sulfuric acid concentration gives no remarkable variation in the recovery of vanadium. Moreover, the high concentration of sulfuric acid will increase the cost of practical production.

Effect of Hydrofluoric Acid Concentration

Hydrofluoric acid was added to destroy the silicate mineral lattice to release more vanadium. The effect of the HF concentration on the vanadium leaching percent was studied with the same condition of the 125 g L^{-1} H_2SO_4, 1 wt% $KMnO_4$, L/S of 5 mL g^{-1} at 90 °C for 4 h. The vanadium leaching percent increased as the

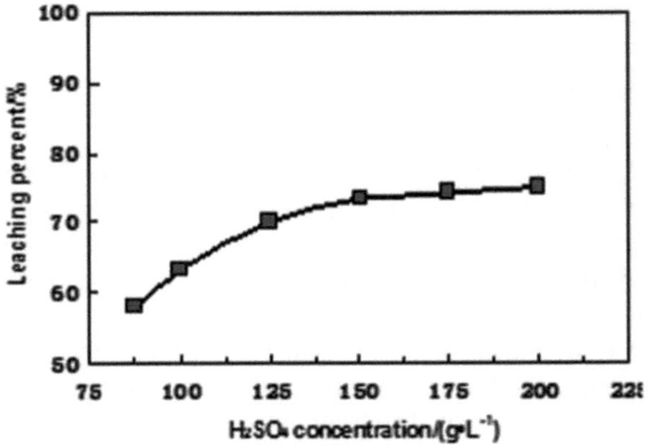

Fig. 2 Effect of H_2SO_4 concentration on percent of vanadium leaching

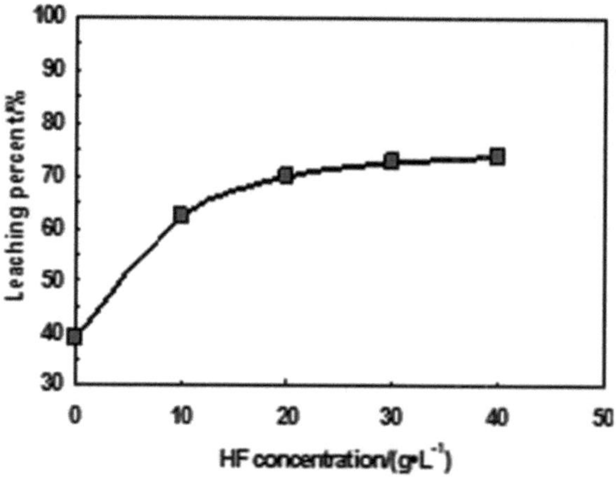

Fig. 3 Effect of HF concentration on percent of vanadium leaching

hydrofluoric acid concentration increased, as shown in Fig. 3. Without HF in the solution, the vanadium leaching ratio is only 39.0% while at a HF concentration of 30 g L^{-1}, the vanadium leaching ratio reached 73.9%. The presence of HF destroys the silicate mineral lattice and the vanadium in this silicate mineral can then be leached with sulfuric acid. However, when the HF concentration exceeds 30 g L^{-1} the vanadium recovery increases slowly as the HF concentration is increased.

Effect of Potassium Permanganate Addition

The destruction of the silicate mineral lattice by hydrofluoric acid exposes vanadium $(trivalence)_{(s)}$ to the mineral surface. To improve the kinetics of the redox-reaction during leaching, $KMnO_4$ was added to oxidize the low valence vanadium to vanadium (tetravalence) or vanadium (pentavalence), which enhances vanadium dissolution in sulfuric acid. The effect of $KMnO_4$ concentration on vanadium leaching percent with the same condition of the 125 g L^{-1} H_2SO_4, 20 g L^{-1} HF, L/S of 5 mL g^{-1} at 90 °C for 4 h, is shown in Fig. 4. We can find that the addition of $KMnO_4$ significantly enhanced the vanadium leaching percent. Without $KMnO_4$, vanadium recovery was 61.3% and with 3 wt% $KMnO_4$, vanadium leaching reached 76.5%. Further addition of $KMnO_4$ does not affect vanadium leaching percent. Clearly, the addition of 3 wt% $KMnO_4$ is sufficient to oxidize low valence vanadium to high valence vanadium.

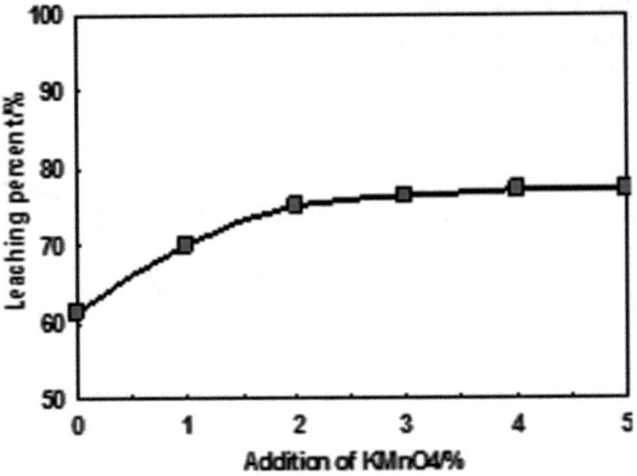

Fig. 4 Effect of addition of KMnO4 on percent of vanadium leaching

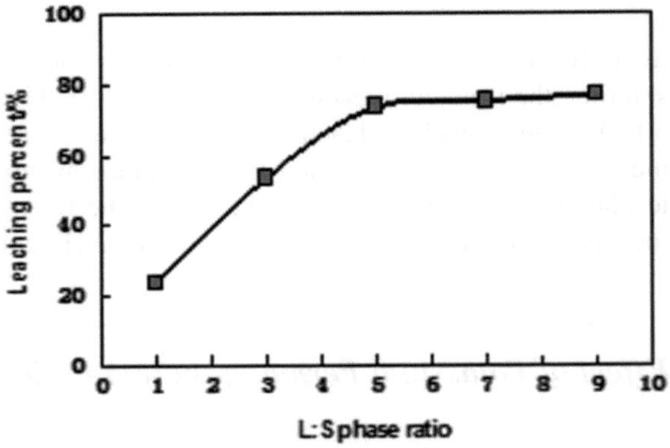

Fig. 5 Effect of reaction L/S phase ratio on percent of vanadium leaching

Effect of the Liquid/Solid Ratio

The effect of the liquid-to-solid ratio on vanadium leaching percent was examined with the same condition of the 125 g L^{-1} H_2SO_4, 20 g L^{-1} HF, 1 wt% $KMnO_4$ at 90 °C for 4 h. As shown in Fig. 5, the vanadium leaching percent increased from 25.1 to 72.4% as the liquid/solid ratio increased. However, when the liquid/solid ratio exceeded 5 mL g^{-1}, the vanadium leaching percent increased slowly as the liquid/solid ratio increased. Practically, a larger liquid/solid ratio means more waste will be produced after extracting vanadium from the leaching solution.

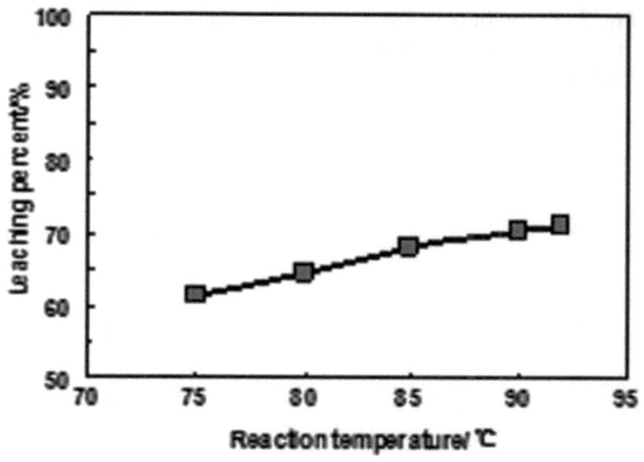

Fig. 6 Effect of reaction temperature on percent of vanadium leaching

Effect of Reaction Temperature

The leaching temperature on effect of vanadium leaching percent was examined under the same condition of the 125 g L^{-1} H_2SO_4, 20 g L^{-1} HF, 1 wt% $KMnO_4$, L/S of 5 mL g^{-1} for 4 h. The vanadium leaching percent versus reaction temperature is shown in Fig. 6.

As shown in Fig. 6, vanadium leaching percent increased with increasing in reaction temperature and leaching percent of vanadium of 70.5% was obtained at a reaction temperature of 90°C.

Effect of Leaching Time

The effect of leaching time on vanadium extraction is shown in Fig. 7 with the same condition of the 125 g L^{-1} H_2SO_4, 20 g L^{-1} HF, 1 wt% $KMnO_4$, L/S of 5 mL g^{-1}

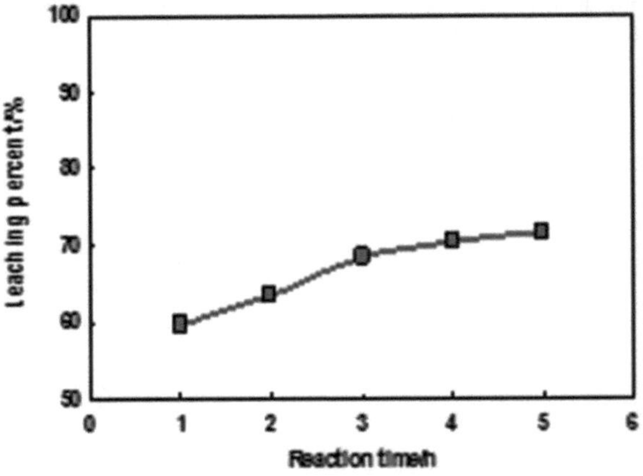

Fig. 7 Effect of reaction time on percent of vanadium leaching

Table 2 Results of confirmatory experiment

No.	Leached percent of vanadium (%)
1	84.8
2	85.7
3	85.0
4	84.7
5	85.2
Average	85.1

at 90 °C. As expected, increased leaching time does lead to increased vanadium extraction. Rapid dissolution of vanadium occurs up to 58.1% within 1 h and then it increases gradually with 70.5% vanadium extracted after 4 h. No further significant change occurs after 4 h.

Confirmatory Experiment

According to the results mentioned above, the optimum conditions of vanadium extraction from extracted vanadium residue of vanadium titanomagnetite by H_2SO_4-HF-$KMnO_4$ system under atmospheric pressure are obtained to granule size of 0.15–0.25 mm, sulphuric acid concentration of 125 g L^{-1}, hydrofluoric acid concentration of 30 g L^{-1}, potassium permanganate dosage of 3 wt%, liquid-to-solid ratio of 5 mL g^{-1}, stirring rev of 500 r min^{-1} at 90 °C for 4 h.

To confirm the vanadium extraction process, five experiments were carried out under these optimum conditions and vanadium leaching percent was over 85%. The results are shown in Table 2.

Conclusions

Based on the discussion above, three conclusions can be drawn:

(1) The vanadium contained in an extracted vanadium residue can be effectively extracted by HF-H_2SO_4-$KMnO_4$ composed leaching system.
(2) HF can destroy the silicate phase which surrounds the vanadium-iron spinel. The addition of $KMnO_4$ can enhance the oxidation of insoluble vanadium (trivalence) to soluble vanadium (tetravalence) or vanadium (pentavalence), thus improve the vanadium leaching percent.
(3) The optimum conditions for extraction vanadium from the extracted vanadium residue are as follows: The extracted vanadium residues are leached with granule size of 0.15–0.25 mm, sulphuric acid concentration of 125 g L^{-1}, hydrofluoric acid concentration of 30 g L^{-1}, potassium permanganate dosage of 3 wt%, liquid-to-solid ratio of 5 mL g^{-1}, stirring rev of 500 r min^{-1} at 90 °C for 4 h, the leaching percent of vanadium is above 85%.

Acknowledgements We acknowledge the National Natural Science Foundation of China (No. 51474115), National Basic Research Program of China (No. 2014CB643404), Yunnan Province Applied Foundation Research Programs (No. 2014FB126) and Analysis and Testing Center of Kunming University of Science and Technology, China.

References

1. Wang M, Xiang X, Zhang L (2008) Effect of vanadium occurrence state on the choice of extracting vanadium technology from stone coal. Rare Met 27:112–115
2. Navarroa R, Guzmana J, Saucedoa I, Revillab J, Guibal E (2007) Vanadium recovery from oil fly ash by leaching, precipitation and solvent extraction processes. Waste Manag 27:425–438
3. Moskalyk RR, Alfantazi AM (2003) Processing of vanadium: a review. Miner Eng 16:793–805
4. Sui Y-L, Guo Y-F, Travyanov AY, Jiang T, Chen F, Qiu G-Z (2016) Reduction roasting–magnetic separation of vanadium tailings in presence of sodium sulfate and its mechanisms. Rare Met 35:954–960
5. Zhang G-Q, Zhang T-A, Zhang Y, Lv G-Z, Liu Y, Liu Z-L (2016) Pressure leaching of converter vanadium slag with waste titanium dioxide. Rare Met 35:576–580
6. Dong Y, Wu X, Yu L, Li L (2007) Fundamental research on vanadium recovering from V-bearing steelmaking slag. Eng Sci 9:63–68 (in Chinese)
7. Lan Y, Liu J (2005) Review of vanadium processing in China. Eng Sci 3:58–62
8. Ye G, Tong X, Lu L (2010) Pretreatment for V-bearing steelmaking slag by beneficiation methods and its effect on followed leaching procedure. Chin J Nonferrous Met 20:2233–2238 (in Chinese)
9. Gahan CS, Cunha ML, Sandstöm Å (2009) Comparative study on different steel slag as neutralizing agent in bioleaching. Hydrometallurgy 95:190–197
10. Yu B, Sun Z, Zhang T, Xian Y, Mu W, Tang H (2014) Non-roasting pressure acid-leaching process of vanadium slag. Rare Met 38:1134–1140 (in Chinese)
11. Voglauer B, Grausam A, Jörgl HP (2004) Reaction- kinetics of the vanadium roast process using steel slag as a secondary raw material. Miner Eng 17:317–321
12. Vitolo S, Seggiani M, Falaschi F (2001) Recovery of vanadium from a previously burned heavy oil fly ash. Hydrometallurgy 62:145–150

13. Feng Q, He D, Zhang G, Ou L, Lu Y (2008) Effect of vanadium oxidation and conversion on vanadium leaching in extraction process of vanadium from stone coal. Chin J Nonferrous Met 17:1348–1352 (in Chinese)
14. He D, Feng Q, Zhang G, Ou L, Lu Y (2007) An environmentally-friendly technology of vanadium extraction from stone coal. Miner Eng 20:1184–1186

Study on Thiosulfate Leaching of Gold by Cycling Barren Solution

Yongbin Yang, Meixiang Lai, Qiang Zhong, Qian Li, Bin Xu and Tao Jiang

Abstract Thiosulfate leaching is widely recognized as one of the most promising alternatives to cyanidation for gold leaching, of which the action of cyclic barren solution is of great importance as far as its commercial application is concerned. In this study, the composition of the leaching solution was characterized and its effect on gold leaching was investigated with regard to cycling thiosulfate barren solution. By direct cycling of barren solution without supplement of any chemical reagent, the gold leaching rate for a gold-bearing calcine decreased noticeably with cycle index, from 71.1% with the fresh solution to 17.8% after four cycles. This is mainly attributed to the decrease in thiosulfate concentration, which turned out to be from 0.5 to 0.06 mol/L. By supplementing thiosulfate and copper into the cycling solutions to the same concentration as the initial fresh solution, associated with addition of a certain amount of sodium sulfite as stabilizer, the gold leaching rate was favorably stabilized as being up to 63.0% during four cycles.

Keywords Thiosulfate leaching of gold · Cycling barren solution · Sodium sulfite Gold recovery

Y. Yang · M. Lai · Q. Zhong (✉) · Q. Li · B. Xu · T. Jiang
School of Minerals Processing and Bioengineering, Central South University, Changsha 410083, China
e-mail: zhongqiang2008csu@163.com

Y. Yang
e-mail: ybyangcsu@126.com

M. Lai
e-mail: LMX17308416672@163.com

Q. Li
e-mail: csuliqian@126.com

B. Xu
e-mail: xuandy16@126.com

T. Jiang
e-mail: jiangtao@csu.edu.cn

G. Azimi et al. (eds.), *Rare Metal Technology 2019*, The Minerals, Metals & Materials Series, https://doi.org/10.1007/978-3-030-05740-4_17

Introduction

Gold is an important strategic material and plays an important role in the development of national economy and social progress. Cyanidation has long been dominant in the gold industry due to its stable chemical properties, simple process and low cost. However, the toxicity of cyanide has shortcomings such as low gold leaching rate and poor leaching of refractory gold ores [1]. Thiosulfate leaching is one of the most promising alternative methods for gold leaching, which has the advantages of being cheap, non-toxic and insensitive to impurities. The research and development of thiosulfate leaching have been motivated by the environmental concerns with the use of cyanide and problems associated with the processing of refractory ores [2]. Karlin and Zubieta [3] first proposed that copper–ammonia complex acts as a catalyst during thiosulfate leaching of gold, which accelerates the dissolution rate of gold by 18–20 times for pure gold.

As a non-cyanide leaching agent, thiosulfate has been extensively studied under bench leaching condition, but the industrial application of thiosulfate leaching is still very rare. The high consumption of thiosulfate and the passivation of gold are two main limitations [4, 5]. Thiosulfate is apt to be oxidized and transformed into other anions, which leads to a large amount of consumption and high production cost. In industrial leaching process, it is necessary to recycle the barren solution to the leaching system to make full use of leaching agents and water. However, because of the instability of thiosulfate and possible accumulation of impurities [6], it is difficult to maintain the gold leaching effect, while recycled solutions are used, as high as leaching with fresh solution. Therefore, technical development on stabilization of the leaching effect with recycled barren solution is of vital significance to widely industrial application of thiosulfate leaching of gold.

In this study, the composition of the barren solution was characterized and its influence on gold leaching effect was investigated with regard to cycling thiosulfate barren solution. Techniques for stabilization of gold leaching effect with recycled barren solutions were thereby researched with much attention.

Experimental

Materials and Reagents

The material used in this paper was roasted calcine from a high-arsenic and high-sulfur gold concentrate processed by two-stage oxidative pretreatment. As a contrast test, cyanide leaching was carried out on the gold calcine, and the gold leaching rate was 77.6%. The grinding fineness was more than 80% which could make them pass through a 200-mesh screen (<0.074 mm), and its chemical composition was shown in Table 1. The content of Au was 36.8 g/t, and the content of Fe, As and S was

Table 1 Chemical composition of the calcine (%)

Constituent	Au[a]	Cu	Fe	Pb	S	As	CaO	MgO	Al_2O_3	SiO_2
Content	36.8	0.20	17.18	0.14	2.724	3.726	3.29	1.20	7.43	34.67

[a] *Unit* g/t

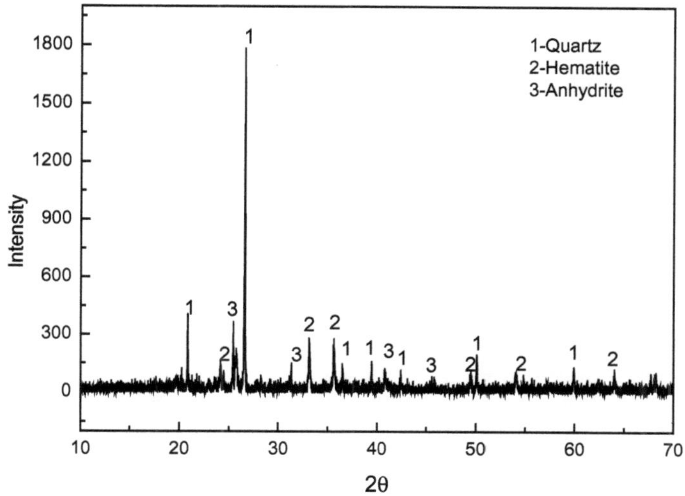

Fig. 1 X-ray diffraction pattern of gold calcine

17.18, 3.726 and 2.724%, respectively. Also, the total content of gangue including SiO_2, CaO, MgO and Al_2O_3 was as high as 46.59%.

The XRD pattern of the calcine (see Fig. 1) showed that quartz and hematite were the major components.

Leaching Experiment and Detection Methods

Four cyclic closed-circuit tests were conducted under laboratory conditions, as shown in Fig. 2. In the thiosulfate leaching test, a 20-g sample was put in a 250-mL breaker equipped with overhead stirrer at a rotating speed of 200 rpm. All experiments were performed at room temperature, the solution pH value was maintained in the range of 9.7–10.2 which was adjusted with 3.0 mol/L NaOH. The concentrations of reagents were: $Na_2S_2O_3$ 0.5 mol/L, $CuSO_4$ 0.03 mol/L, $NH_3 \cdot H_2O$ 1.5 mol/L. The liquid–solid rate was 3:1, and the leaching time was 8 h. The reagents used in this study were all analytically pure, and deionized water was used throughout all experiments. When a gold leaching test was completed, the slurry was filtered by a vacuum filter and the pregnant solution was analyzed to determine gold leaching rate and the chemical

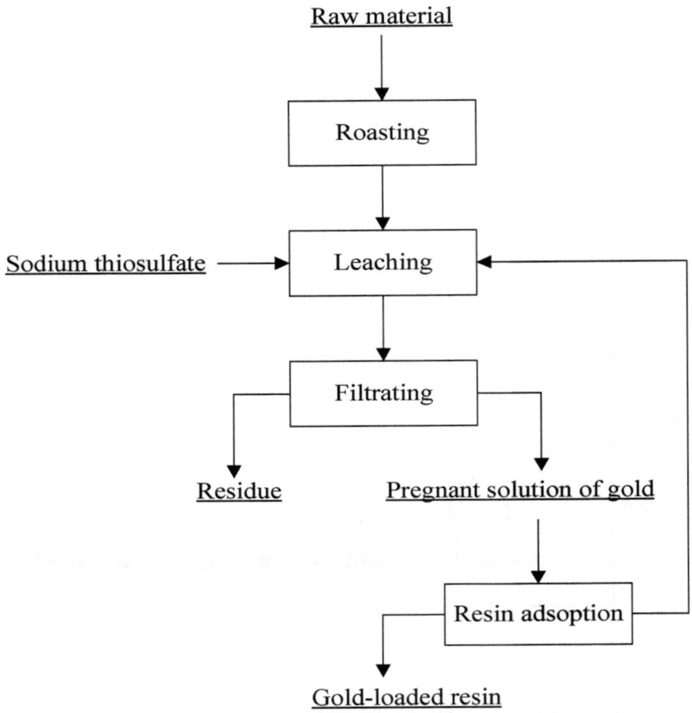

Fig. 2 Thiosulfate cyclic leaching flow sheet

composition of the solution. The pregnant solution was then processed by resin adsorption or copper powder replacement for recovery of gold and cycling of barren solutions.

The concentrations of metallic elements in the solutions were determined by ICP-OES, and the thiosulfate concentration was determined by iodometric method [7]. In order to eliminate the effect of the cupric tetra-amine complex and sulfite on iodine titration, a certain amount of EDTA-2Na and formaldehyde (50%, volume fraction) were added prior to the titration with the indicator Vitex [8].

Leaching rate of gold in cyclic leaching

$$\eta = \frac{C_N V_N - C_{N-1} V_{N-1}}{m_0}$$

C_N—The concentration of gold leachate in the Nth cycle, mg/L
V_N—The volume of the Nth cyclic leachate, L
C_{N-1}—The concentration of barren solution in the $N - 1$th cycle, mg/L
V_{N-1}—The volume of the $N - 1$th cyclic leachate, L
m_0—The amount of gold contained in the sample, mg.

Results and Discussion

To investigate gold leaching with cyclic barren solution, the compositions of leaching solution before and after resin adsorption were analyzed for the leaching with the initial fresh leachate.

Under initial leaching conditions ($Na_2S_2O_3$: 0.5 mol/L, $CuSO_4$: 0.03 mol/L, $NH_3 \cdot H_2O$: 1.5 mol/L), the concentration of gold and the concentration of thiosulfate in the pregnant solution were 6.532 mg/L and 0.360 mol/L (in Table 2), from which the calculated gold leaching rate turned out to be 71.1%, and the consumption rate of thiosulfate was 28.0%. The thiosulfate in the barren solution was recycled, which will make full use of leaching agents to save cost. The main role of copper ions in thiosulfate leaching was to catalyze the dissolution of gold, resulting in its consumption. The concentration of metallic ions in the initial leachate was very low and changed little before and after resin adsorption. In cyclic leaching of thiosulfate barren solution, metallic ions may gradually accumulated and have detrimental effect on gold leaching. Therefore, it is necessary to study accumulation of metallic ions and stability of the leaching effect with recycled barren solution.

Cyclic Leaching of Thiosulfate Barren Solution

Under initial leaching conditions ($Na_2S_2O_3$: 0.5 mol/L, $CuSO_4$: 0.03 mol/L, $NH_3 \cdot H_2O$: 1.5 mol/L), the gold leaching rate was 71.1% in the initial fresh leaching. By direct cycling of barren solution without supplement of any chemical reagent, the pH of leaching process was adjusted to about 10 with ammonia water. The results of cyclic leaching were shown in Fig. 3. The addition of $NH_3 \cdot H_2O$ can be used not only as a pH adjustment agent but also as supplement to the concentration of NH_4^+ in the leaching system. When the barren solution does not added any reagent for cyclic leaching, the gold leaching rate decreased noticeably with cycle index from 71.1% with the fresh solution to 17.8% after four cycles. This was mainly attributed to the decrease in thiosulfate solution which turned out to be 0.5 mol/L to 0.06 mol/L. Furthermore, the concentration of copper as a catalyst also gradually decreased with

Table 2 Compositions of leaching solution before and after resin adsorption

Component	Au mg L^{-1}	Cu mg L^{-1}	Fe mg L^{-1}	Pb mg L^{-1}	Zn mg L^{-1}	$S_2O_3^{2-}$ M
Initial leachate	–	1920.0	–	–	–	0.50
Pregnant solution	6.532	1275.2	6.355	4.935	3.177	0.360
Barren solution	0.1965	712.5	6.078	5.028	3.005	0.349

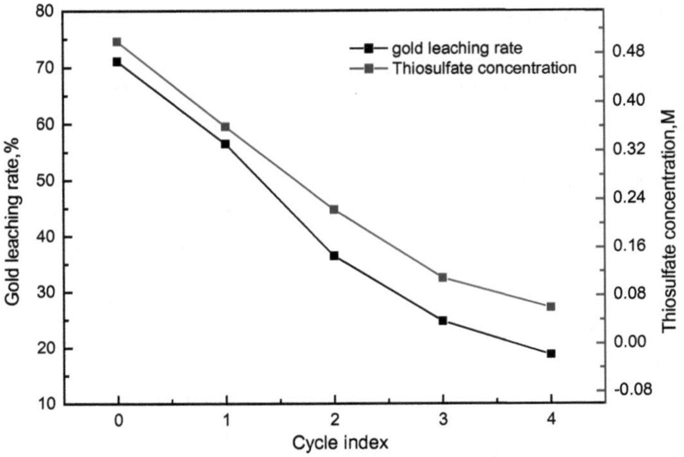

Fig. 3 Gold leaching rate and thiosulfate concentration versus cycle index

the cycle index (see Fig. 4), which detrimental to thiosulfate leaching of gold. Consequently, copper powder replacement may be a good choice for gold recovery because the dissolved in thiosulfate barren solution of copper ions can be reused in cyclic leaching.

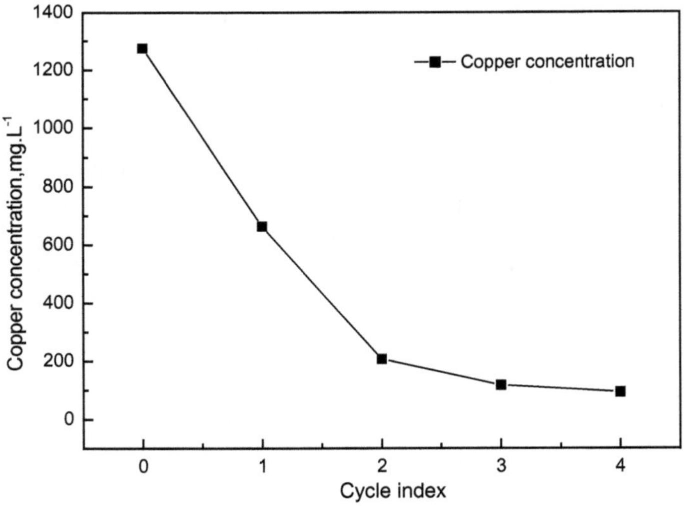

Fig. 4 Copper concentration versus cycle index

Effect of Sodium Sulfite on Cyclic Leaching

A certain amount of sodium sulfite was added to the initial fresh leaching and cyclic leaching of thiosulfate barren solution. On the one hand, thiosulfate can be prevented from decomposing into sulfite and sulfur; on the other hand, it can promote the decomposition of polysulfate and reduce the loss of $S_2O_3{}^{2-}$ [9]. According to Eqs. (1) and (2), $SO_3{}^{2-}$ can effectively reduce the decomposition of $S_2O_3{}^{2-}$ and maintain the stability of $S_2O_3{}^{2-}$ in the solution. Since $SO_3{}^{2-}$ will be oxidized in the thiosulfate leaching system, too much $SO_3{}^{2-}$ will consume a large amount of dissolved oxygen into $SO_4{}^{2-}$. Gudkov et al. [10] also proposed that appropriate $SO_3{}^{2-}$ can effectively reduce the decomposition of $S_2O_3{}^{2-}$, maintain the stability of $S_2O_3{}^{2-}$ in the leaching system and promote the dissolution of gold.

$$2S_2O_3^{2-} + 0.5O_2 + H_2O \rightarrow S_4O_6^{2-} + 2OH^- \quad (1)$$

$$SO_3^{2-} + S_4O_6^{2-} \rightarrow S_3O_6^{2-} + S_2O_3^{2-} \quad (2)$$

Adding 0.3 mol/L sodium sulfite as stabilizer under optimal conditions, and replenishing thiosulfate and cupric into the cycling solutions to the same concentration as the initial fresh solution, the results of cyclic leaching were shown in Fig. 5. In initial fresh leaching, the gold leaching rate was 71.8% and the consumption rate of thiosulfate was 26.4%. Compared to Fig. 3, sodium sulfite in initial leaching not only improved the gold leaching rate but also reduced the consumption of thiosulfate. Meanwhile, the gold leaching rate was favorably stabilized as being up to 63.0% during four cycles. Compared with the initial leaching, the gold leaching rate of cycling barren solution was slightly reduced, but remained stable. Moreover, the concentration of thiosulfate in barren solution was stable, which greatly saved the amounts of leaching agent. It is of great significance for industrial application of thiosulfate leaching of gold to stabilization of the leaching effect with recycled barren solution.

The effects of cycle index on the concentration of metallic ions under above cyclic leaching conditions are shown in Fig. 6. After the fourth-cycle leaching, the concentrations of iron, lead and zinc were 6.355, 4.262 and 5.871 mg/L, respectively. Metallic ions tend to accumulate slowly, but the concentration of metal ions was very low during cyclic leaching compared with that of those contained in calcine. If impurity ions accumulate in large amounts, it will have adverse effects on gold leaching rate, gold recovery rate, and so on. There was no obvious accumulation of metallic ions, and the barren solution can be returned to reuse. The effect of metallic ions on gold leaching depends on its concentration in solution, and even some low concentration of metallic ions can facilitate the leaching of gold [11].

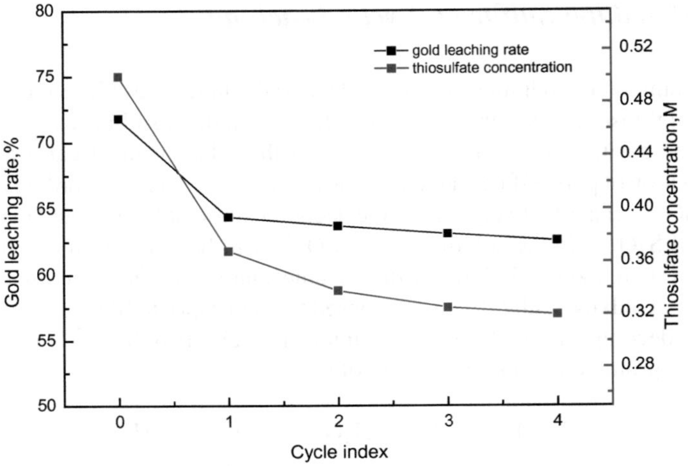

Fig. 5 Gold leaching rate and thiosulfate concentration versus cyclic index

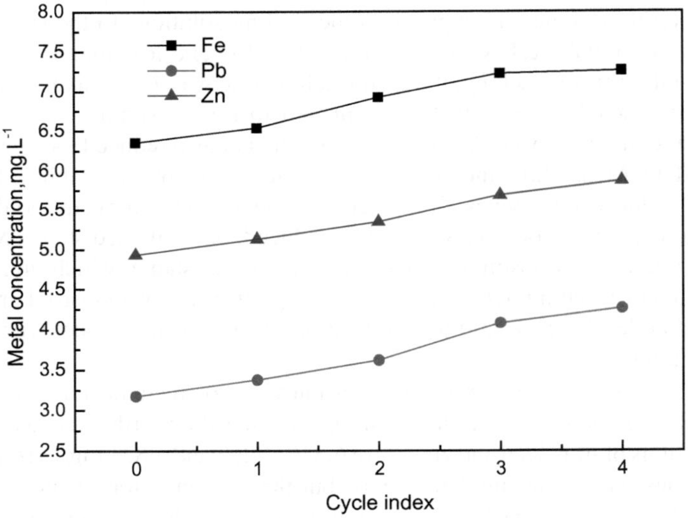

Fig. 6 Concentration of metallic ions versus cycle index

Table 3 Effect of metallic ions on the stability of thiosulfate

Cation type	Concentration of cation, mg/L	Initial concentration of $S_2O_3^{2-}$, mol/L	Final concentration of $S_2O_3^{2-}$, mol/L	Loss rate of $S_2O_3^{2-}$, %
Zn^{2+}	30	0.1875	0.1856	1.01
Pb^{2+}	30	0.1875	0.1825	2.67
Fe^{3+}	30	0.1875	0.1833	2.24

The influence of equal amount of metallic ions on the stability of thiosulfate was studied. As we can seen from Table 3, when metallic ions Zn^{2+}, Pb^{2+}, Fe^{3+} 30 mg/L were added in the sodium thiosulfate, the loss rate of $S_2O_3^{2-}$ was small and even had little effect. The concentration of metallic ions in barren solution was almost no influence on thiosulfate leaching of gold. This also shown that the thiosulfate itself has a strong ability to resist foreign cationic interference.

Gold Recovery by Resin and Copper

Figures 7 and 8 show the effect of resin and copper on gold recovery rate and copper concentration. The recovery of gold used a strong base anion resin 2 g/L and copper–gold mass ratio of 300. While the recovery time was long enough, the gold recovery rate by copper and resin was basically the same up to 98% (see Fig. 7). The concentration of copper decreased gradually in the process of resin recovery, while the opposite was true by copper replacement (see Fig. 8). The presence of appropriate copper ions has a positive effect on the thiosulfate leaching of gold. Copper may be a good choice for gold recovery, because the dissolved in thiosulfate solution of copper ions can be reused in the barren solution. Moreover, both copper and cuprous ions can significantly enhance the semi-reaction of gold deposition [12]. Copper powder as a reducing agent to gold recovery was a promising method for the replacement copper cycled into the solution, which has certain potential utilization value in production practice [13, 14]. It is the future research direction to reduce copper precipitation and copper recycling and realize the cyclic leaching of thiosulfate barren solution.

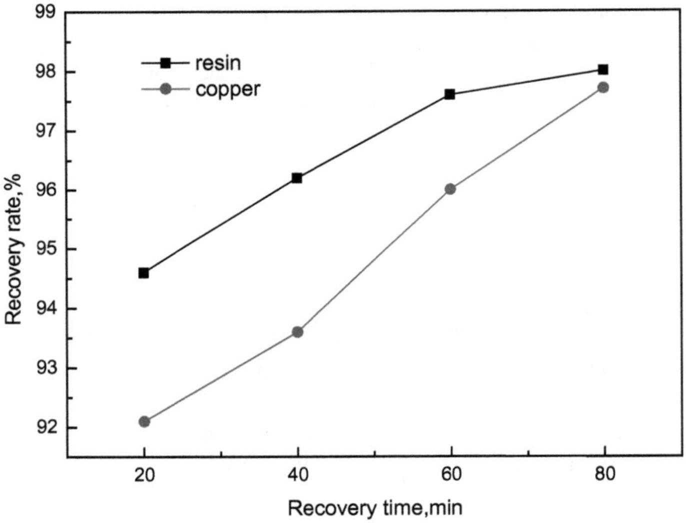

Fig. 7 Gold recovery by resin and copper

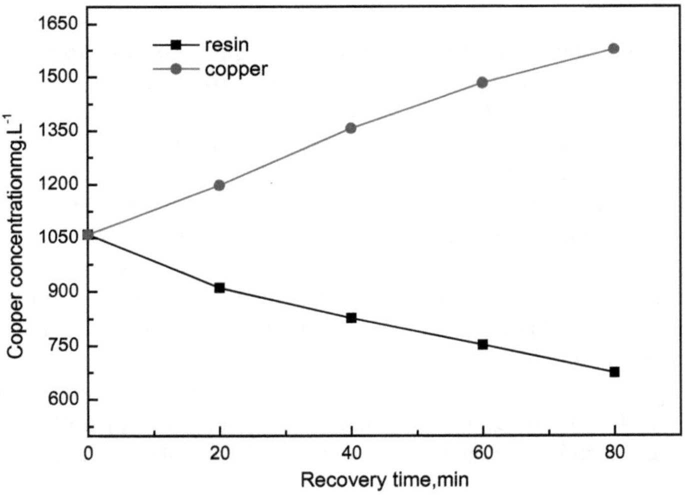

Fig. 8 Effect of resin and copper on copper concentration

Conclusions

The thiosulfate cyclic leaching of gold by barren solution was investigated. The following conclusions can be drawn:

(1) The leaching rate of gold decreases greatly by direct cyclic leaching with barren solution, which is attributed to a gradual decrease in concentrations of thiosulfate and copper.
(2) Adding 0.3 mol/L sodium sulfite as stabilizer under optimal conditions, and replenishing thiosulfate and cupric into the cycling solutions to the same concentration as the initial fresh solution. In initial fresh leaching, it not only improves gold leaching rate but also reduces thiosulfate consumption. Meanwhile, the gold leaching rate was favorably stabilized as being up to 63.0% during four cycles.
(3) The recovery rate of gold is basically the same by resin and copper powder, but the concentration of copper in barren solution was much different. Copper may be a good choice for gold recovery, because the dissolved in barren solution of copper ions can be reused in cyclic leaching.

References

1. La Brooy SR, Linge HG, Walker GS (1994) Review of gold extraction from ores. Miner Eng 7(10):1213–1241
2. Abbruzzese C, Fornari P, Massidda R, Veglio F, Ubaldini S (1995) Thiosulphate leaching for gold hydrometallurgy. Hydrometallurgy 39(1–3):265–276
3. Karlin KD, Zubiet J (1983) Copper coordination chemistry: biochemical & inorganic perspectives. Adenine Press
4. Jeffrey MI, Watling K, Hope GA, Woods R (2008) Identification of surface species that inhibit and passivate thiosulfate leaching of gold. Miner Eng 21(6):443–452
5. Senanayake G (2004) Analysis of reaction kinetics, speciation and mechanism of gold leaching and thiosulfate oxidation by ammoniacal copper (II) solutions. Hydrometallurgy 75(1):55–75
6. Feng D, Deventer JSJ (2010) The effect of iron contaminants on thiosulphate leaching of gold. Miner Eng 23(5):399–406
7. Gu XC (2008) Inorganic chemical reaction mechanism. Chemical Industry Press, Beijing
8. Liu XL, Xu B, Yang YB, Li Q, Jiang T (2017) Effect of galena on thiosulfate leaching of gold. Hydrometallurgy 171:157–164
9. Chu CK, Breuer PL, Jeffrey MI (2003) The impact of thiosulfate oxidation products on the oxidation of gold in ammonia thiosulfate solution. Miner Eng 16(3):265–271
10. Gudkov AS, Zhuchkov IA, Mineev GG (2010) Mechanism and kinetics of sulfite-thiosulfate dissolution of gold. Russ J Non-Ferrous Met 51(5):393–397
11. Feng D, Deventer JSJ (2004) The role of heavy metallic ions in gold dissolution in the ammoniacal thiosulphate system. Hydrometallurgy 64(3):231–246
12. Choo WL, Jeffrey MI (2004) An electrochemical study of copper cementation of gold(I) thiosulfate. Hydrometallurgy 71(3):351–362

13. Karavasteva M (2010) Kinetics and deposit morphology of gold cemented on magnesium, aluminum, zinc, iron and copper from ammonium thiosulfate-ammonia solutions. Hydrometallurgy 104(1):119–122
14. Arima H, Fujita T, Yen WT (2002) Gold cementation from ammonium thiosulfate solution by zinc, copper and aluminium powders. Mater Trans 43(3):485–493

Part V
Poster Session

Leaching of Tellurium and Bismuth from the Dashuigou Tellurium Deposit in H_2SO_4 and $FeCl_3$ Media

Li-Xiong Shao, Jiang Diao, Liang Liu and Bing Xie

Abstract In order to recover the tellurium and bismuth resources within the Dashuigou tellurium deposit in Sichuan province, China, the leaching of these metals from the raw ore in H_2SO_4 and $FeCl_3$ media was investigated in this work. The effect of H_2SO_4 concentration, Fe^{3+} concentration, liquid/solid ratio, leaching time and leaching temperature was studied. It was found that 99.17% of Te and 92.82% of Bi were extracted into leaching solution under the optimum conditions: H_2SO_4 concentration of 150 g/L, Fe^{3+} concentration of 60 g/L, liquid/solid ratio of 6 ml/g, leaching time of 60 min and leaching temperature of 90 °C. Furthermore, the sulfur element in raw ore was reduced into elementary substance. There is no production of H_2S, which is a positive indicator for environmental considerations.

Keywords Tellurium · Bismuth · Leaching · Sulfuric acid · Ferric chloride

Introduction

Tellurium and bismuth are strategic metals, which are critical for applications in modern industry and national defense. Due to their excellent thermal, optical and electrical properties, tellurium and tellurium-containing compounds are extensively applied in various industries such as steel and glass manufacturing, petroleum refining, solar panels, sensor production, medicine and rechargeable battery manufacturing [1, 2]. Similarly, bismuth and bismuth-containing compounds are also widely used in metallurgy, alloys, medicines, chemicals, and electronics [3, 4]. Nowadays, most tellurium and bismuth are recovered from the by-product of copper or lead anode slime [5, 6]. The traditional process of recovering tellurium from anode slime through soda ash roasting followed by hot water leaching, purification, neutraliza-

L.-X. Shao · J. Diao (✉) · L. Liu · B. Xie
Chongqing Key Laboratory of Vanadium-Titanium Metallurgy and Advanced Materials, College of Materials Science and Engineering, Chongqing University, Chongqing 400044, People's Republic of China
e-mail: diaojiang@163.com

G. Azimi et al. (eds.), *Rare Metal Technology 2019*, The Minerals, Metals & Materials Series, https://doi.org/10.1007/978-3-030-05740-4_18

tion, incineration, alkali-dissolving, and electrolysis. However, due to the reasons like long flow and complexity of traditional industrial methods, the recovery rate is low (60–70%) [7]. The conventional extraction process of bismuth from its sulfide mineral bismuth glance (bismuthinite) is a pyrometallurgy process (>1200 °C) that involves the reduction of bismuth sulfide (Bi_2S_3) to metallic bismuth, which can cause serious problems such as environmental pollution and high energy consumption [5]. Thus, many methods have been put forward to improve the extraction ratio of tellurium and bismuth, including acid leaching [8–10], pressure sulfate leaching [11], pressure alkali leaching [12, 13] and bioleaching [14, 15]. Although the bioleaching is economic and energy-efficient, this method is still limited since the leaching time is longer and the leaching scale is very small. The pressure leaching is also greatly limited in industrial applications due to the limitations of leaching equipment. Besides the bio-leaching and pressure leaching, the acid leaching also has attracted attention. Some researchers [9, 10] proposed the method to recover the tellurium and bismuth by leaching in hydrochloric acid or nitric acid media. However, due to the severe volatilization of hydrochloric and nitric acid, sulfuric acid was put forward to be a good lixiviant for the extraction of metal. Ha et al. [16] indicated that the extraction of bismuth from copper converter dust is very effective since the bismuth in the converter dust is in the form of the oxide. Hait et al. [17] showed that the extraction of tellurium from copper anode slime always remained lower level when sulfuric acid media was used for leaching of the anode slime without any additive. It indicates that sulfuric acid plays a limited role in the leaching of sulfide ore or more complex ore. Hait et al. [17] also presented that the extraction of tellurium by sulfuric acid leaching with the addition of MnO_2 is more efficient than that without any additive. The recovery of tellurium can further increase to 66% with the increase of temperature to 80 °C. Other additions besides MnO_2 were also found to improve the extraction of metal, including H_2O_2, $NaClO_3$, and $FeCl_3$ [18–20]. Among the additions, ferric chloride can be recycled by chlorination; therefore, the acidic ferric chloride leaching has certain advantages for the sulfide ore or more complex ore.

With the discovery of the tellurium deposit in Sichuan province of China [21], the recovery of tellurium and bismuth from the tellurium deposit has attracted significant interest due to the increasing use of tellurium throughout China. This tellurium deposit contains a remarkably high content of low-grade tellurium sulfides [22]. However, this tellurium deposit has not been utilized to recover the tellurium and bismuth, due to the lack of economically viable technology. Based on the considerations mentioned above, the recovery of tellurium and bismuth from tellurium deposit in sulfuric acid media with the addition of ferric chloride is proposed in this work. The effects of various parameters on the dissolution of tellurium and bismuth have been systematically investigated. The mineralogical phases of the leaching residue were also studied.

Table 1 Chemical composition of raw tellurium ore (wt%)

Te	Bi	S	Fe_2O_3	SiO_2	CaO	MgO	K_2O	Na_2O	Al_2O_3	TiO_2	Cu
1.52	2.99	13.67	36.82	3.60	16.51	8.07	0.32	0.16	1.97	0.11	0.09

Experimental

Materials

The tellurium ore used in the experiments was the raw ore from Dashuigou tellurium deposit in Sichuan province, China. The chemical analysis of the raw tellurium ore is shown in Table 1. The chemical analysis showed that the raw tellurium ore contained low-grade tellurium and bismuth and a higher content of S and Fe. The XRD analysis of tellurium ore revealed two main solid phases: pyrite (FeS_2) and dolomite ($CaCO_3{\cdot}MgCO_3$), which is presented in Fig. 1. No tellurium or bismuth solid phases were detected by XRD, possibly due to the poor crystallization property of its compounds [7]. According to Refs. [21, 22], the solid phases of the ore consisted mainly of pyrite (FeS_2), pyrrhotite (FeS), dolomite ($CaCO_3{\cdot}MgCO_3$), daphyllite (Bi_2Te_2S), telluric bismuth (BiTe) and quartz (SiO_2).

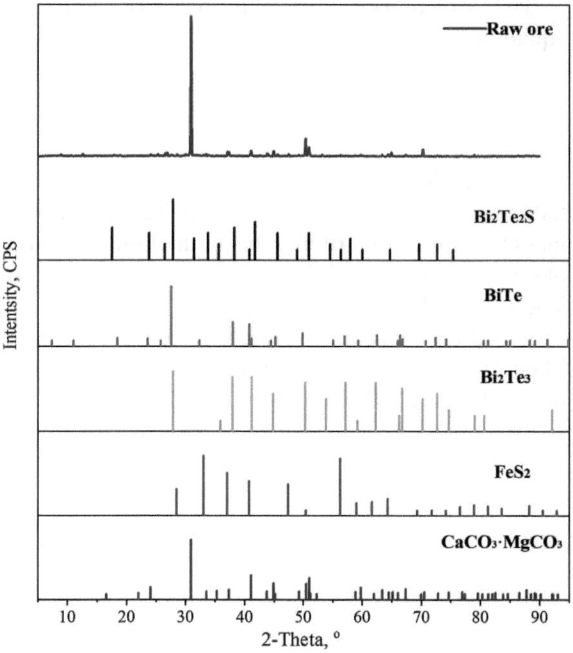

Fig. 1 XRD pattern of the raw ore

The H_2SO_4 and $FeCl_3$ media solution was prepared with H_2SO_4 (>98.0%, mass) and analytical reagent $FeCl_3$ (>98.0%, mass), which were provided by Chengdu Kelong Chemical Reagent Factory.

Experimental Procedures of Leaching

Media solutions with Fe^{3+} concentration of 100 g/L were prepared with H_2SO_4 and analytical reagent $FeCl_3$ by varying H_2SO_4 concentration from 0 to 250 g/L. After optimum H_2SO_4 concentration was obtained, other media solutions with optimum H_2SO_4 concentration were prepared by varying Fe^{3+} concentration from 0 to 150 g/L. The tellurium ore was ground to –200 mesh and then dried at 105 °C for 24 h to remove the moisture. One gram of raw ore was used for each test. Firstly, the ore was fed into a three-necked flask. Then, a definite amount of H_2SO_4 and $FeCl_3$ media solution was added. Finally, the reaction mixture was immediately placed in a commercial water bath pot to keep the temperature at the designed temperature without stirring. Different leaching conditions were tested, and finally, concentrations of the tellurium and bismuth in solution were performed by using inductively coupled plasma-atomic emission spectrometry (ICP-AES, Optima 4300DV, Perkin Elmer, USA). The influential factors, including H_2SO_4 concentration, Fe^{3+} concentration, liquid/solid ratio, leaching time and leaching temperature, were investigated.

The extraction ratios of tellurium and bismuth were calculated using the following equation:

$$\eta = \frac{V_m C_t}{m_0 w_0} \times 100\% \tag{1}$$

where η is the extraction ratio of tellurium or bismuth, %. m_0 is the weight of the tellurium ore every leaching experiment, g. In this work, the value of m_0 is 1. w_0 is the mass fraction of tellurium or bismuth of the tellurium ore, %. C_t is the concentration of tellurium or bismuth in leach liquor at the final moment, g/L. V_m is the volume of leach liquor, L.

Based on the optimum leaching conditions, the leaching residue was obtained. Then, the mineralogical analysis of leaching residue was carried out by X-ray diffraction analysis (XRD, RIGAKUD/MAX 2500PC).

Results and Discussion

Effect of H_2SO_4 Concentration

Figure 2 shows the effect of the H_2SO_4 concentration on the extraction ratio of tellurium and bismuth. With the increase of the H_2SO_4 concentration in the media solution, the extraction ratios of tellurium and bismuth increase and then plateau. When H_2SO_4 concentration increases from 0 to 100 g/L, the extraction ratios of tellurium and bismuth increase from 1.20 to 97.83% and from 14.60 to 89.64%, respectively. Any further increase in H_2SO_4 concentration results in a marginal change in the extraction of tellurium and bismuth. During the leaching process, sulfuric acid acts as an acid source to increase the H^+ activity, which can increase both the oxidizing potential of the Fe^{3+} ions and the potential for oxidation of tellurium and bismuth of the ore to corresponding ions [10]. In addition, sulfuric acid can provide SO_4^{2-} to remove the Ca^{2+} in the solution for the following Na_2SO_3/SO_2 reduction to obtain elemental tellurium with high purity. Based on these results, H_2SO_4 concentration of 100 g/L is considered as the optimum H_2SO_4 concentration for subsequent leaching experiments. However, considering the H_2SO_4 is used to dissolve dolomite and remove sufficiently calcium ions by producing calcium sulfate. Thus, H_2SO_4 concentration of 150 g/L is considered as the optimum H_2SO_4 concentration for subsequent leaching experiments.

Effect of Fe^{3+} Concentration

During the leaching process, the iron trivalent acts as oxidation, which is beneficial to the leaching of tellurium and bismuth from the raw ore (Eqs. 2 and 3). As shown in

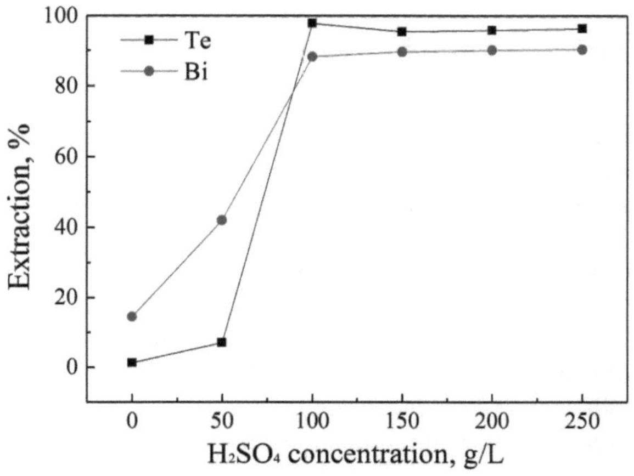

Fig. 2 Effect of H_2SO_4 concentration on tellurium and bismuth extractions (Fe^{3+} concentration of 100 g/L, liquid/solid ratio of 6 ml/g, leaching time of 120 min and leaching temperature of 80 °C)

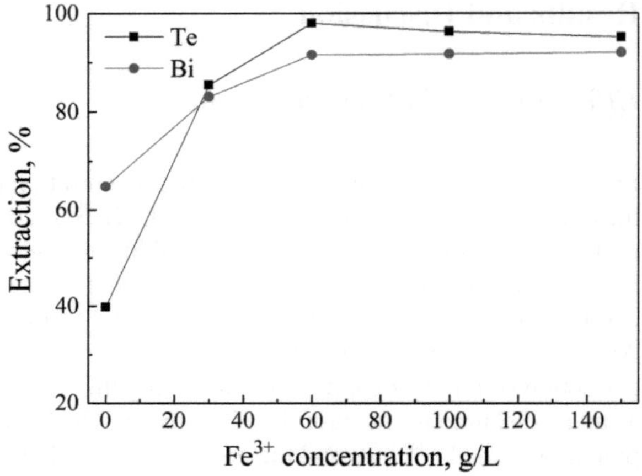

Fig. 3 Effect of Fe^{3+} concentration on tellurium and bismuth extractions (H_2SO_4 concentration of 150 g/L, liquid/solid ratio of 6 ml/g, leaching time of 120 min and leaching temperature of 80 °C)

Fig. 3, with the increase of Fe^{3+} concentration in the media solution from 0 to 60 g/L, the extraction ratios of tellurium and bismuth increase from 39.80 to 97.99% and from 64.83 to 91.53%, respectively. Any further increase in Fe^{3+} concentration results in a marginal change in the extraction of tellurium and bismuth. The recovery of tellurium remained almost lower level in sulfuric acid without Fe^{3+}. Similar behavior has been reported by Hait et al. [17] in sulfuric acid leaching of Te-bearing anode slime. It indicates that sulfuric acid leaching with the iron trivalent is a good method for the recovery of tellurium and bismuth from the tellurium ore. Hence, Fe^{3+} concentration of 60 g/L is considered as the optimum Fe^{3+} concentration for subsequent leaching experiments.

$$Bi_2Te_2S(\mathrm{s}) + 14Fe^{3+}(\mathrm{aq}) = 2Bi^{3+}(\mathrm{aq}) + 2Te^{4+}(\mathrm{aq}) + 14Fe^{2+}(\mathrm{aq}) + S(\mathrm{s}) \quad (2)$$

$$BiTe(\mathrm{s}) + 7Fe^{3+}(\mathrm{aq}) = Bi^{3+}(\mathrm{aq}) + Te^{4+}(\mathrm{aq}) + 7Fe^{2+}(\mathrm{aq}) \quad (3)$$

Effect of Liquid/Solid Ratio

Figure 4 presents the effect of the liquid/solid ratio on the extraction ratio of tellurium and bismuth. It can be seen that the extraction ratio of tellurium slightly increases and shows a value of more than 94% with the increase of the liquid/solid ratio from 4 to 10 ml/g. When the liquid/solid ratio reaches 6 ml/g, the extraction ratio of tellurium can reach a higher value, 97.99%. However, the extraction ratio of bismuth increases from 81.22 to 91.53% by increasing the liquid/solid ratio from 4 to 6 ml/g. Any further increase in liquid/solid ratio results in a marginal fluctuation in the extraction of bismuth. The higher liquid/solid ratio is beneficial to the extraction of bismuth. It indicates that the increase of the liquid/solid ratio would promote the mass transfer process at the solid–liquid interface [18, 23]. Therefore, in order to achieve a higher

Fig. 4 Effect of liquid/solid ratio on tellurium and bismuth extractions (H_2SO_4 concentration of 150 g/L, Fe^{3+} concentration of 60 g/L, leaching time of 120 min and leaching temperature of 80 °C)

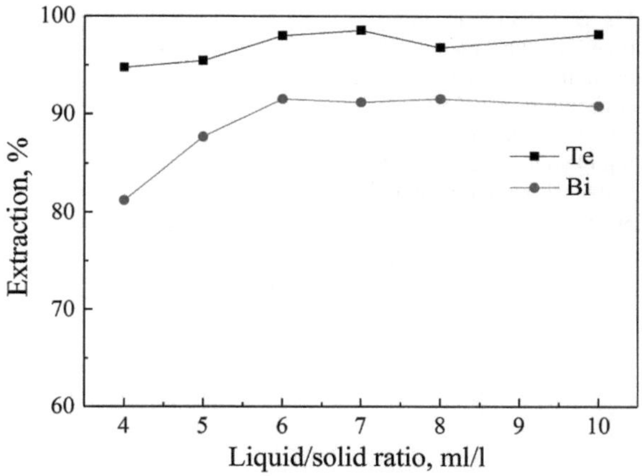

extraction ratio of tellurium and bismuth and make full use of the media solution, the optimal liquid/solid ratio of 6 ml/g was selected for the following experiment.

Effect of Leaching Time

The effect of the leaching time on the extraction ratio of tellurium and bismuth is shown in Fig. 5. With the prolongation of leaching time, the tellurium extraction is almost independent of leaching time and shows a value of more than 96.70%. In comparison, the extraction of bismuth increases slowly with time, and 90.98% bismuth can be obtained after leaching for 60 min. Further increase in leaching time gives no further benefit for the leaching of bismuth. In the initial stage of leaching experiment, the reaction of sulfuric acid and dolomite generates a large amount of carbon dioxide, which would promote the mixing of media solution and solid. Thus, it further promotes mass transfer process at the solid–liquid interface to improve the reaction efficiency. Consequently, leaching time of 60 min was considered optimal for the following experiments.

Effect of Leaching Temperature

The results summarized in Fig. 6 show an increase of leaching efficiency of tellurium and bismuth with the increase of leaching time from 25 to 98 °C. The extraction ratios of tellurium and bismuth increase slowly with increasing the leaching temperature and reach a maximum value at 90 °C. When the leaching temperature increases from 25 to 90 °C, the extraction ratios of tellurium and bismuth increase from 92.01 to 99.17% and from 82.99 to 92.82%, respectively. It indicates high leaching tempera-

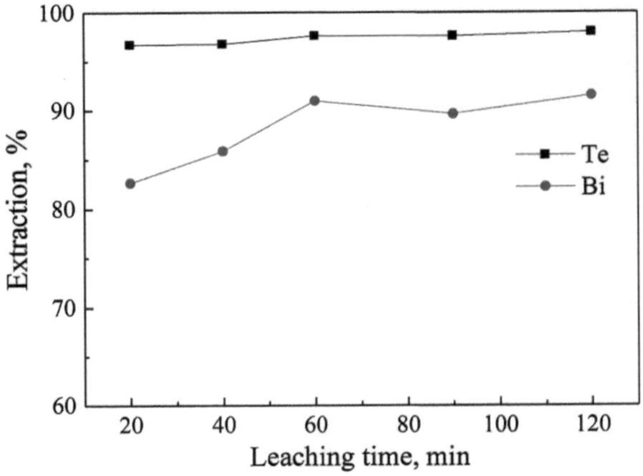

Fig. 5 Effect of leaching time on tellurium and bismuth extractions (H_2SO_4 concentration of 150 g/L, Fe^{3+} concentration of 60 g/L, liquid/solid ratio of 6 ml/g and leaching temperature of 80 °C)

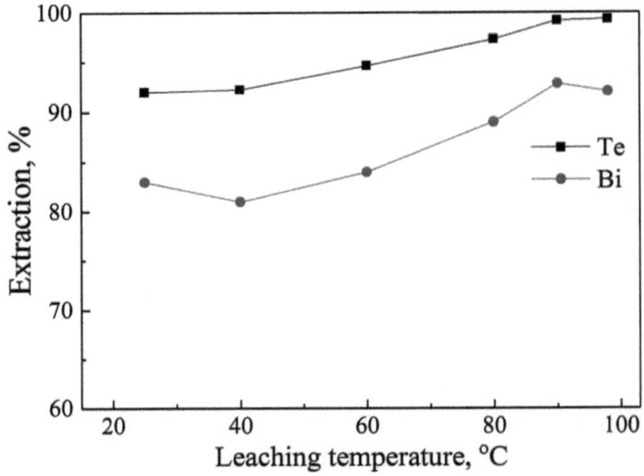

Fig. 6 Effect of leaching temperature on tellurium and bismuth extractions (H_2SO_4 concentration of 150 g/L, Fe^{3+} concentration of 60 g/L, liquid/solid ratio of 6 ml/g and leaching time 60 min)

ture will give a higher extraction ratio of tellurium and bismuth. Therefore, leaching temperature of 90 °C was selected as the optimum leaching temperature in the current study.

The Mineralogical Phases of the Leaching Residue

Based on these results, the optimum leaching conditions can be obtained. Figure 7 shows the XRD spectrum of leaching resides under the optimal leaching conditions. The phases of leaching reside are calcium sulfate ($CaSO_4$) and sulfur (S). The above results indicate that the sulfuric acid with ferric chloride is a very effective lixiviant in dissolving tellurium and bismuth from tellurium ore. Ca^{2+} also can be removed into leaching reside. Thus, the leach liquor will directly be used for the following

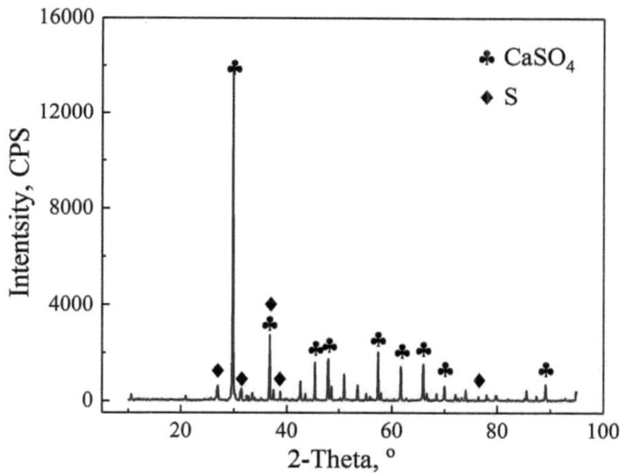

Fig. 7 XRD pattern of the leaching residue

Na_2SO_3/SO_2 reduction to obtain elemental tellurium with high purity. Furthermore, the sulfur in raw ore is oxidized to the elemental sulfur into leaching reside, which can avoid the pollution of an environment from H_2S since it can be produced in sulfuric acid leaching process without ferric chloride.

Conclusion

The leaching of tellurium and bismuth in sulfuric acid media with the addition of ferric chloride is proposed in this work to recover the tellurium and bismuth of Dashuigou tellurium deposit in Sichuan province, China. The influences factors on the extraction of tellurium and bismuth were studied, including H_2SO_4 concentration, Fe^{3+} concentration, liquid/solid ratio, leaching time and leaching temperature. It was found that 99.17% of Te and 92.82% of Bi were extracted into leaching solution under optimum conditions: H_2SO_4 concentration of 150 g/L, Fe^{3+} concentration of 60 g/L, solid/liquid ratio of 6 ml/g, leaching time of 60 min and leaching temperature of 90 °C. Moreover, the sulfur element in raw ore was reduced to elementary substance into leaching reside, without production of H_2S, which indicated that this process is environmentally friendly. In addition, the Ca^{2+} in the leach liquor can be removed by producing calcium sulfate. Thus, the leach liquor will directly be used for the following Na_2SO_3/SO_2 reduction to obtain elemental with high purity.

References

1. Turner RJ, Borghese R, Zannoni D (2012) Microbial processing of tellurium as a tool in biotechnology. Biotechnol Adv 30(5):954
2. Ba LA, Doering M, Jamier V, Jacob C (2010) Tellurium: an element with great biological potency and potential. Org Biomol Chem 8(19):4203
3. Bothwell JM, Krabbe SW, Mohan RS (2011) Applications of bismuth(III) compounds in organic synthesis. Tetrahedron 40(9):4649–4707
4. Keogan DM, Griffith DM (2014) Current and potential applications of bismuth-based drugs. Molecules 19(9):15258–15297
5. Chen Y, Liao T, Li GB, Chen BZ, Shi X (2012) Recovery of bismuth and arsenic from copper smelter flue dusts after copper and zinc extraction. Miner Eng 39(12):23–28
6. Makuei FM, Senanayake G (2018) Extraction of tellurium from lead and copper bearing feed materials and interim metallurgical products—A short review. Miner Eng 115:79–87
7. Guo X, Xu Z, Li D, Tian QH, Xu RZ, Zhang Z (2017) Recovery of tellurium from high tellurium-bearing materials by alkaline sulfide leaching followed by sodium sulfite precipitation. Hydrometallurgy 171:355–361
8. Fan XX, Xing WD, Dong HG, Zhao JC, Wu YD, Li BJ, Tong WF, Wu XF (2013) Factors research on the influence of leaching rate of nickel and cobalt from waste superalloys with sulfuric acid. Int J Nonferr Metall 2(2):63–67
9. Jdid EA, Elamari K, Blazy P, Hakkou R (1996) Acid and oxidizing leaching of copper refinery anodic slimes in hexafluorosilicic acid and nitric acid media. Sep Sci 31(4):569–577
10. Li XJ, Yang HY, Jin ZN, Tong LL, Xiao FX (2017) Selenium leaching from copper anode slimes using a nitric acid-sulfuric acid mixture. Metallurgist 61(3–4):348–356
11. Zhang BY, Wang JK, Peng JH (2007) Research on removing tellurium from copper anode slime by process of acid leaching under pressure. Nonferr Metal 4:27–29
12. Liu W, Yang T, Zhang D, Chen L, Liu Y (2014) Pretreatment of copper anode slime with alkaline pressure oxidative leaching. Int J Miner Process 128(5):48–54
13. Rhee KI, Lee CK, Ha YC, Jeong GJ, Kim HS, Sohn HJ (1999) Tellurium recovery from cemented tellurium with minimum waste disposal. Hydrometallurgy 53(53):189–201
14. Rajwade JM, Paknikar KM (2003) Bio-reduction of tellurium to elemental tellurium by *Pseudomonas mendocino* MCMB-180 and its practical application. Hydrometallurgy 71(3):243
15. Guo YF, Zhang N, Li DC, Tang FM, Deng TL (2012) Tellurium recovery from the unique tellurium ores. Adv Mater Res 549(549):1060–1063
16. Ha TK, Kwon BH, Park KS, Mohapatra D (2015) Selective leaching and recovery of bismuth as Bi_2O_3 from copper smelter converter dust. Sep Purif Technol 142:116–122
17. Hait J, Jana RK, Kumar V, Sanyal SK (2002) Some studies on sulfuric acid leaching of anode slime with additives. Ind Eng Chem Res 41(25):6593–6599
18. Aydogan S, Aras A, Canbazoglu M (2005) Dissolution kinetics of sphalerite in acidic ferric chloride leaching. Chem Eng J 114(1–3):67–72
19. Simon FG, Holm O, Berger W (2013) Resource recovery from urban stock, the example of cadmium and tellurium from thin film module recycling. Waste Manage 33(4):942–947
20. Fthenakis VM, Wang W (2006) Extraction and separation of Cd and Te from cadmium telluride photovoltaic manufacturing scrap. Prostate 14(4):363–371
21. Yin JZ, Chen YC, Zhou JX, Yang BC, Zhang Y, Mao JW (1994) Mineralogical study of the world's first independent tellurium deposit. Bull Min, Petrol Geochem 13(3):153–155
22. Mao JW, Chen YC (2010) Geology and geochemistry of the Dashuigou tellurium deposit, western Sichuan, China. Int Geo Rev 37(6):526–546
23. Zhang S, Li T, Zhu B, Zhu Z (2005) Gas-liquid mass transfer in three-phase mechanical agitated reactor. J Chem Ind Eng 56(2):200–226

New Dissolution Process of Iridium to Hydrochloric Acid

Yuto Kobayashi, Shota Yamada and Takashi Nagai

Abstract Iridium is used in various industrial products, one such application being a thin film catalysis of an oxygen-evolving anode. When iridium and iridium oxide are recovered and purified, they must be dissolved to aqueous solution due to them being chemically stable the dissolution in acid is difficult. Currently, it is dissolved with hydrochloric acid to which chlorine gas is injected but this leads to problems of safety and large environmental load. It is necessary to improve dissolving behavior of iridium in acid. In this study, a method to dissolve iridium to hydrochloric acid is developed. By mixture of metal Ir powder and $CaCO_3$ was heated at 973 K or higher temperature, Ca_2IrO_4 and Ca_4IrO_6, complex oxides were synthesized. Iridium in the oxides formed can easily be dissolved in hydrochloric acid.

Keywords Iridium · Recycling · Electrode · Hydrochloric acid

Introduction

The platinum group metals (PGM) consists of platinum (Pt), palladium (Pd), rhodium (Rh), ruthenium (Ru), iridium (Ir) and osmium (Os). Iridium is used for electrodes, chemical catalysts and spark plugs and is widely used in industry. Production volume is small at approximately 4 tons per year and supply is unstable because resources are in limited areas such as South Africa and Russia. Since it is produced from low-grade ore, enormous energy is required for smelting and the environmental burden is large. For these reasons, recycling iridium from used products is critical.

For recovery and purification of iridium, wet methods such as solvent extraction and precipitation separation are used. Therefore, it is necessary to dissolve the iridium in solution. However, since it is a chemically stable substance, dissolution is very difficult. It is insoluble in concentrated nitric acid and concentrated sulfuric acid, even

Y. Kobayashi · S. Yamada · T. Nagai (✉)
Department of Mechanical Science and Engineering, The University of Chiba Institute of Technology, 2-17-1 Tudanuma, Narasino, Chiba 275-0016, Japan
e-mail: takashi.nagai@it-chiba.ac.jp

G. Azimi et al. (eds.), *Rare Metal Technology 2019*, The Minerals, Metals & Materials Series, https://doi.org/10.1007/978-3-030-05740-4_19

aqua regia (a mixture of nitric and hydrochloric acid) hardly dissolves it in the bulk state. For dissolution, acids containing a strong oxidizing agent such as hydrochloric acid aerated with chlorine gas are used. These are problematic in terms of safety due to high corrosivity and toxicity. Therefore, if improvement of the acid solubility of iridium and dissolution with relatively safe hydrochloric acid becomes possible, a great improvement in safety in the recovery and purification of the element can be expected.

To improve the acid solubility of iridium, we focused on a method for synthesized complex oxide. In Kasuya's report, when Pt and Li salt is mixed and heated, Li_2PtO_3 is synthesized and it can easily be dissolved in hydrochloric acid [1]. Therefore, in this study, Ir-containing complex oxide was synthesized and its solubility in hydrochloric acid was investigated.

Experiment

A sample was prepared by mixing Ir (powder) and oxide or carbonate powder for synthesizing complex oxide. (Oxide or carbonate; $BaCO_3$, BiO_2, $CaCO_3$, Li_2CO_3, Na_2CO_3, NbO, PbO, $SrCO_3$, Y_2O_3, ZrO_2.) Also, a sample of only Ir was prepared. The prepared sample was placed in an alumina crucible and heated in an electric resistance furnace. After heating, the sample was crushed and analyzed by X-ray diffraction (XRD) to identify the generated phase.

To investigate whether the heated sample was soluble in hydrochloric acid 0.1 g of the heated sample and 20 mL of hydrochloric acid were placed in a beaker. Watch glass was placed on the beaker and dissolved while heating with a hot plate (Fig. 2). The dissolution temperature was 353 K and the dissolution time was 3 h.

After dissolving, if there was a residue in the solution, it was filtered. Thereafter, the solution was diluted to 100 mL with pure water. This solution was analyzed using ICP-OES, and the concentration of Ir was measured. If the concentration of Ir exceeded the range of the calibration curve, it was diluted and analyzed again. The dissolution rate of Ir was calculated from the following formula:

$$S = \frac{\rho \times r \times 10^{-4}}{m_{\mathrm{Ir}}'} \times 100 \tag{1}$$

where,

S Dissolution rate of Ir (mass%)
ρ Measured concentration of Ir (ppm)
r Dilution ratio
m_{Ir}' Mass of Ir contained in the dissolved sample (g).

$$m_{\mathrm{Ir}}' = m \times \frac{m_{\mathrm{Ir}}}{m_{\mathrm{Ir}} \times K_{\mathrm{Ir}} \times m_{\mathrm{reagent}} \times K_{\mathrm{reagent}}} \tag{2}$$

where,

m	Mass of dissolved sample (g)
m_{Ir}	Mass of mixed Ir (g)
$m_{reagent}$	Mass of mixed reagent (g)
K_{Ir}	Chemical formula weight of Ir after heating per atomic weight of Ir (g/mol)
$K_{reagent}$	Chemical formula weight of reagent after heating per molecular weight of reagent (g/mol).

Results and Discussion

Ir and IrO_2 were dissolved with concentrated hydrochloric acid and the solubility was investigated. The experimental conditions and dissolution rates of Ir and IrO_2 are listed in Table 1. IrO_2 was prepared by heating Ir under the conditions of 1273 K, 6 h. The dissolution rate of IrO_2 was calculated assuming that Ir became IrO_2. The dissolution rates of Ir and IrO_2 were 0.054, 0.014 mass%, and dissolution with hydrochloric acid was very difficult.

The dissolution rate of Ir in sample and synthesization of complex oxide in heating experiment at 973 K are summarized in Table 2. In the cases of Bi_2O_3, $CaCO_3$, Li_2CO_3, Na_2CO_3 and PbO, the complex oxide was synthesized completely. However, the dissolution rate of Ir in sample was small except $CaCO_3$. Those of the others were also small.

In the Ir–$CaCO_3$ system, the generated phase and solubility at the heating temperature were investigated. Samples of Ir + $CaCO_3$ were heated at temperatures 1573 K, heating time was 1 h. Then, it was dissolved with concentrated hydrochloric acid. The dissolution rate was calculated assuming that Ir became IrO_2 and $CaCO_3$ became CaO.

The generated phase at the heating temperature of 1573 K, heating time 1 h is shown in Fig. 1. A peak of Ca_4IrO_6 was formed at a heating temperature of 1573 K and heating time 1 h at which the dissolution rate was high. The dissolution rate was 100 mass%, almost all of the iridium was dissolved.

Through experiments, it was found that Ir can easily be dissolved with hydrochloric acid when the complex oxide Ca_4IrO_6 is prepared. Further studies into the dissolution behavior of Ir-containing complex oxides are needed to find the effective factor for improving solubility.

Table 1 Dissolution rate of Ir and IrO_2

Sample	HCl concentration (c/mol L^{-1})	Dissolution rate S (mass%)
Ir	approx. 12	0.054
IrO_2		0.014

Table 2 Dissolution rate of Ir in sample

Sample	Dissolution rate S (mass%)	Complex oxide
Ir + $BaCO_3$	48	△
Ir + Bi_2O_3	4.2	○
Ir + $CaCO_3$	68	○
Ir + Li_2CO_3	19	○
Ir + Na_2CO_3	0.78	○
Ir + NbO	0.12	×
Ir + PbO	0.52	○
Ir + $SrCO_3$	36	△
Ir + Y_2O_3	0.0	×
Ir + ZrO_2	0.039	×

Circle Single phase of complex oxide
Triangle Complex oxide and other phases
Multiplication sign No complex oxides

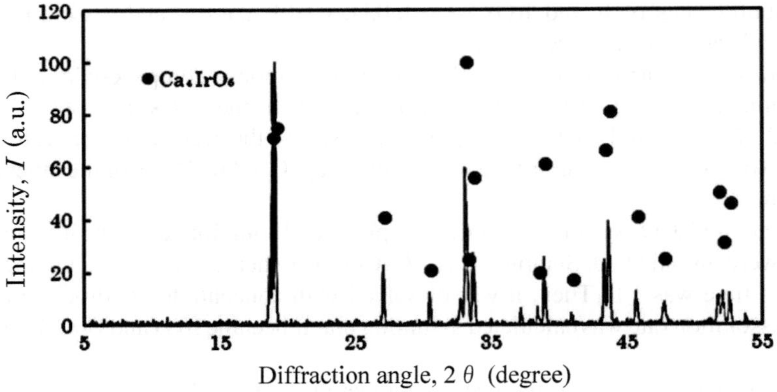

Fig. 1 X-ray diffraction patterns of synthesized Ca_4IrO_6

Reference

1. Kasuya R, Miki T, Morikawa H, Tai Y (2014) Development of new dissolution process of platinum via double oxides. J Japan Inst Met Mater 78(7):242–249

Research on the Carbothermic Reduction Procedure of $SrSO_4$ with Carbon

Si-ming Chen, Dong-ping Duan and Xing-wu Zou

Abstract Carbothermic reduction of celestite is the first procedure of universal process for preparing $SrCO_3$. However, only one species as SrS without SrO can be found in reduction product of $SrSO_4$ by solid carbon at the reduction temperature below 1273 K, and less effect of factors such as C/O ratio and reaction temperature on the reduction of $SrSO_4$ is found. An investigation on reduction of $SrSO_4$ is carried out. The experiment result reveals that the reduction of $SrSO_4$ is a reduction–decomposition process: $SrSO_4$ is first reduced to $SrSO_3$, and then, the $SrSO_3$ was decomposed into $SrSO_4$ and SrS below 1273 K or SrO and SO_2 above 1273 K. Kinetic analysis of the reduction of $SrSO_4$ reveals that the reduction of $SrSO_4$ is controlled by step of interface reaction, and its activation energy is 105.4 kJ/mol.

Keywords Celestite · Carbon-containing pellet · Reduction mechanism
Thermodynamics · Kinetics

S. Chen (✉) · D. Duan
Key Laboratory of Green Process Engineering,
Institute of Process Engineering, Chinese Academy of Sciences, Beijing, China
e-mail: chensiming323171@163.com

S. Chen · X. Zou
Key Laboratory of Comprehensive and Highly Efficient Utilization of Salt Lake Resources,
Qinghai Institute of Salt Lakes, Chinese Academy of Sciences, Xining, Qinghai, China

S. Chen · X. Zou
Key Laboratory of Salt Lake Resources Chemistry of Qinghai Province, Qinghai Engineering and Technology Research Center of Comprehensive Utilization of Salt Lake Resources, Xining, Qinghai, China

S. Chen · D. Duan · X. Zou
University of Chinese Academy of Sciences, Beijing, China

G. Azimi et al. (eds.), *Rare Metal Technology 2019*, The Minerals, Metals & Materials Series, https://doi.org/10.1007/978-3-030-05740-4_20

Introduction

Strontium compounds are important raw materials for many new functional materials [1–3]. However, because of the lower solubility of $SrSO_4$ in water and acidic or alkaline solution [4], it is important to transform celestite, which is the primary strontium mineral in nature, into SrS at first, which is used to prepare $SrCO_3$ at second. As a result, the reduction efficiency of celestite is one of the factors that limits the yield of $SrCO_3$. At present, a rotary kiln with the disadvantage of lower reduction efficiency, higher energy consumption and lower utilization ratio of powder ore is used for carbothermal reduction of celestite in strontium plants. Various effects such as catalyst and activation method are tested to improve its reduction efficiency, but no effective promotion has been achieved. Large dust generation and serious clogging at the inlet of kiln have hindered its application. Furthermore, only lump ore can be used in rotary kiln. Thus, a large amount of low-grade celestite ore from Qinghai, China, cannot be utilized for preparing $SrCO_3$ in rotary kiln.

Rotary hearth furnace (RHF) was applied to treat vanadium–titanium magnetite, iron concentrate, among others. In RHF process, ore powder is mixed with solid reducing agents and pressed into carbon-containing pellets and then placed at the bottom of the RHF for reducing. At sufficient contact between ore powder and reducing agent, the reduction rate is rapid. In addition, lower loss of reactant can be achieved since the reactants and the bottom of the furnace are relatively static. It can effectively solve the problem of serious clogging at inlet of kiln and hardly using concentrate powder. Therefore, it can be predicted that the RHF has lower material consumption and higher reduction efficiency. In a previous work, celestite carbon-containing pellet was reduced in RHF [5]. Better reduction efficiency of celestite was achieved. But less report has been published on the mechanism and kinetic analysis of $SrSO_4$ reduction process.

In this paper, a research on analyzing the thermodynamics of $SrSO_4$ was carried out at first. Then, the mechanism of $SrSO_4$ reduction was studied. The third step was focused on determining the kinetic parameters of $SrSO_4$ reduction process.

Experiment Procedure

Materials

$SrSO_4$ (A.R.) was purchased from Aladdin. Graphite (A.R.) was purchased from Sinopharm Chemical Reagent Co., Ltd. Particle sizes of $SrSO_4$ and graphite are listed in Table 1. All other reagents used in the experiments are A.R. Water used in the experiments is deionized water.

Table 1 Particle sizes of $SrSO_4$ and graphite (wt%)

Grain size	Size distribution	
	$SrSO_4$	Graphite
>600 μm	1.4	0.4
600–212 μm	3.1	8.8
<212 μm	95.5	90.8

Experimental Procedure

The reduction process of celestite carbon-containing pellet was carried out as follows [5]: a celestite carbon-containing pellet was manufactured by mixing of $SrSO_4$ and graphite. And it was added into RHF for reducing for a certain reaction time at specified temperature with nitrogen protection. Keeping the nitrogen protected until the temperature of RHF reached room temperature, reduction product of $SrSO_4$ was collected to determine the reduction ratio of $SrSO_4$.

Experiment Characterization and Analysis

The thermodynamic data of $SrSO_4$ are calculated by HSC (version 6.0). The content of soluble sulfide in reduction product of $SrSO_4$ was determined according to the method described in a previous study [6]: The reduction product of $SrSO_4$ was leached in micro-boiling water with solid-to-liquid ratio as 10 g/L for 1 h. The sulfide in leachate was determined by I_2–$Na_2S_2O_3$ titration method.

The content of sulfide in the reduction product of $SrSO_4$ n_{sulfide}(mol) can be calculated as follows:

$$n_{\text{sulfide}} = \frac{C_{Na_2S_2O_3} \times (V_0 - V_1)}{2} \tag{1}$$

where $C_{Na_2S_2O_3}$ is the concentration of $Na_2S_2O_3$ (mol/L), V_0 is the volume of $Na_2S_2O_3$ which is used to titrate blank solution (L), and V_0 is the volume of $Na_2S_2O_3$ which is used to titrate solution which is added leachate (L).

The reduction ratio of $SrSO_4$ α_{SrSo_4}(%) can be calculated as follows:

$$\alpha_{SrSo_4} = \frac{n_{\text{sulfide}} M_S}{m_{SrSo_4} \omega_S} \times 100 \tag{2}$$

where n_{sulfide} is the content of sulfide in reduction $SrSO_4$ (mol), ω_S is the molar mass of S (32 g/mol), m_{SrSo_4} is the mass of $SrSO_4$ in reactant, and M_S is the mass fraction of S in $SrSO_4$ (0.174%).

Results and Discussion

Thermodynamic Analysis

In the carbothermal reduction process, the amount of C, CO, CO_2, O_2 and H_2O (g) in the reactant affects the reduction of $SrSO_4$. The possible reactions are as follows:

$$2SrSO_4 = 2SrO + O_2 + 2SO_2 \quad (3)$$

$$SrS + 4CO_2 = SrCO_3 + SO_2 + 3CO \quad (4)$$

$$SrS + H_2O(g) + CO_2 = SrCO_3 + H_2S \quad (5)$$

$$C + CO_2 = 2CO \quad (6)$$

$$SrSO_4 + 4CO = SrS + 4CO_2 \quad (7)$$

$$SrSO_4 + 2C = SrS + 2CO_2 \quad (8)$$

$$SrSO_4 + 4C = SrS + 4CO \quad (9)$$

$$SrS + 2O_2 = SrSO_4 \quad (10)$$

$$2C + O_2 = 2CO \quad (11)$$

$$2SrSO_4 + 2C + O_2 = 2SrCO_3 + 2SO_2 \quad (12)$$

$$SrS + 2O_2 = SrO + SO_2 \quad (13)$$

$$SrSO_4 + C = SrO + SO_2 + CO \quad (14)$$

The calculated standard Gibbs free energy from the above equation in the range of 873–1873 K is shown in Fig. 1. It reveals that the $\Delta_r G^\ominus$ calculated for reactions (3) and (4) maintain positive values in the range of studied temperature, indicating that these reactions are difficult to carry out when temperature is changing. The $\Delta_r G^\ominus$ calculated for reaction (5) have a large positive shift with the increase of reaction temperature. It reveals that newly generated SrS does not react with H_2O (g) and CO_2 to form $SrCO_3$ at higher temperature. All of $\Delta_r G^\ominus$ calculated for reactions (7)–(9) are negative, inferring that both solid-based reducing agent C and gas-based reducing agent CO can effectively reduce $SrSO_4$. And C has a better reduction capability for $SrSO_4$ since $\Delta_r G^\ominus$ calculated for reactions (8) and (9) are less than that of reaction (7) as a temperature above 973 K. Comparing the reactions (9) and (10), it indicates that the increase of C/O molar ratio (the ratio of C molar of solid carbon and O molar of $SrSO_4$) can accelerate the reduction process of $SrSO_4$. Reactions (10) and (13) infer that SrO cannot be oxidized from SrS. $\Delta_r G^\ominus$ of reaction (14) are negative when the reaction temperature is higher than 1259 K, indicating that SrO can form only under the condition of lower C/O molar ratio and higher temperature above 1259 K. By comparing with Eqs. (10)–(12), it is known that O_2 may influence the reduction process of $SrSO_4$. To minimize the effect, the content of O_2 should be controlled in at lower levels.

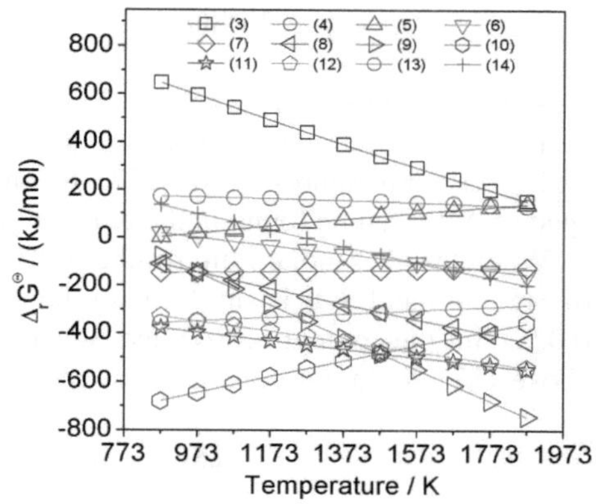

Fig. 1 Comparison of calculated $\Delta_r G^\Theta$ over a temperature range from 873 to 1873 K for Eqs. (3)–(14)

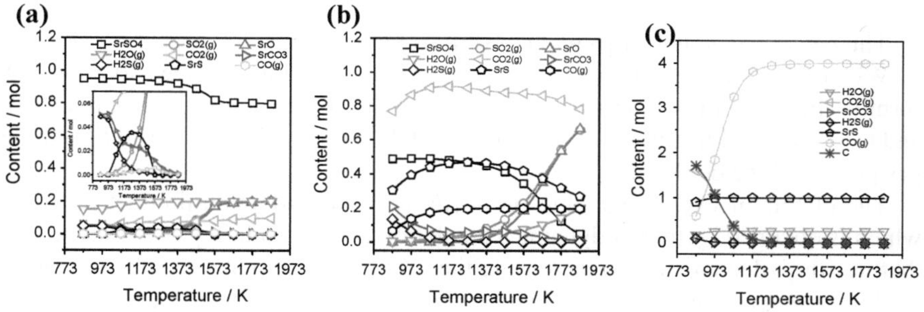

Fig. 2 Thermodynamic calculation results of equilibrium compositions over a temperature range from 873 to 1873 K in five cases: **a** C/O = 0.025; **b** C/O = 0.25; **c** C/O = 1, respectively

It is shown that the C/O molar ratio in the reducing agent and temperature have effects on the reduction efficiency of $SrSO_4$. Thermodynamic calculation results of equilibrium compositions with different C/O molar ratio and temperature were simulated. When C/O molar ratio is 0.025, as shown in Fig. 2a, a large amount of $SrSO_4$ exist in the reduction product. A small amount of $SrCO_3$ can form below 973 K. SrS begins to generate at 973 K. Reverse V-type plot of content of SrS against temperature can be obtained in the range of 973–1573 K, and the maximum value appears at 1273 K. SrS does not form as the temperature exceeds 1573 K. SrO begins to form at 1259 K, and its content increases rapidly with temperature promoted and reaches a stable value as temperature exceeds 1573 K. It is a remarkable fact that changing trend of content of SO_2 is similar to SrO. It begins to generate at 1073 K, and its content promotes rapidly and finally stabilizes at above 1573 K. In addition, the content of SrO and SO_2 is very close under that condition. As a result, the reaction under this condition is as follows:

$$SrSO_4 + C = SrS + CO_2 \tag{15}$$

$$2SrSO_4 + C = 2SrO + 2SO_2 + 2CO_2 \quad (16)$$

When the temperature is below 1273 K, only reaction (15) occurs, while both reaction (15) and reaction (16) occur simultaneously at a temperature of above 1273 K. At higher temperatures, above 1573 K, only reaction (16) occurs. Thus, only one species, SrO, can be found in the reaction product of $SrSO_4$ at that condition.

When the C/O molar ratio reaches 0.25, as shown in Fig. 2b, the changing trend of SrS, SrO and SO_2 is similar with the results obtained using C/O molar ratio as 0.025. But the content of SrO and SO_2 is still increasing as temperature increases above 1573 K. At this condition, the following reactions occur:

$$2SO_2 + 4CO = S_2(g) + 4CO_2 \quad (17)$$

$$3S_2(g) + 4SrO = 4SrS + SO_2 \quad (18)$$

The newly generated $S_2(g)$ will convert SrO into SrS.

When the C/O molar ratio reaches 1.0, as shown in Fig. 2c, few SrO and SO_2 form in the temperature range is studied. It can also be explained by reactions (17) and (18). SrO will transform into SrS completely with ample $S_2(g)$. And SO_2 in the gas phase will be depleted as well.

The effects of C/O molar ratio on n_{SrO}/n_{SrS} in the reduction product of $SrSO_4$ which is obtained by thermodynamic calculation are shown in Fig. 3. It can be seen that n_{SrO}/n_{SrS} decreases with increasing the C/O molar ratio at a fixed temperature. And if n_{SrO}/n_{SrS} is defined as 0.01 when no SrO is generated, the value of C/O molar ratio which can only generate SrO will increase gradually with increasing the reaction temperature, from 0.075 at 1173 K to 0.675 at 1873 K.

As it is mentioned above, reduction process of $SrSO_4$ has variable mechanism at different reaction temperatures. The reaction mechanism summarized by researches is as shown in Table 2.

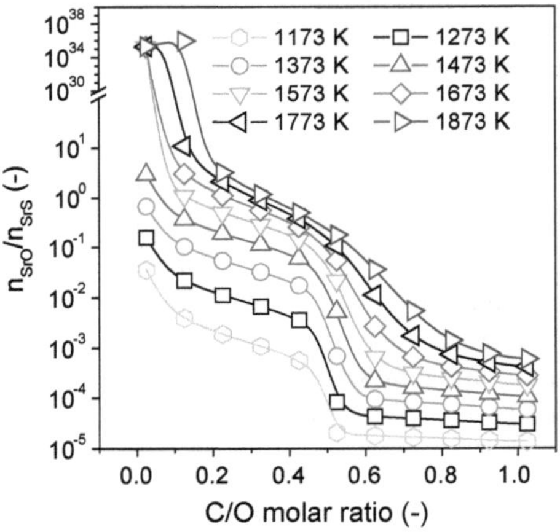

Fig. 3 Relationship between C/O molar ratio of reaction agent and n_{SrO}/n_{SrS} ratio of reduction product under the condition of changing reaction temperature from 1173 to 1873 K

Table 2 Summary of mechanisms of sulfide reduction

Mechanism 1 [7]	Mechanism 2 [8]	Mechanism 3 [9]
$SrSO_4 + C = SrSO_3 + CO$ $SrSO_3 = SrO + SO_2$ $SO_2 + C = S_2(g) + CO$ $S_2(g) + SrO = SrS + SO_2$	$SrSO_4 + C = SrSO_3 + CO$ $SrSO_3 + C = [SrSO_2]^* + CO$ $[SrSO_2]^* + C = [SrSO]^* + CO$ $[SrSO]^* + C = SrS + CO$	$SrSO_4 + C = SrSO_3 + CO$ $4SrSO_3 = 3SrSO_4 + SrS$ $4SrSO_3 = 3SrSO_4 + SrS$ (LT[a]) Or $SrSO_3 = SrO + SO_2$ (HT[b]) $SO_2 + C = S_2(g) + CO$ $S_2(g) + SrO = SrS + SO_2$

[a]LT, low temperature, $T < 1273$ K
[b]HT, high temperature, $T > 1273$ K
Note $[SrSO_2]^*$ and $[SrSO]^*$ are intermediate reduction products of $SrSO_3$, respectively

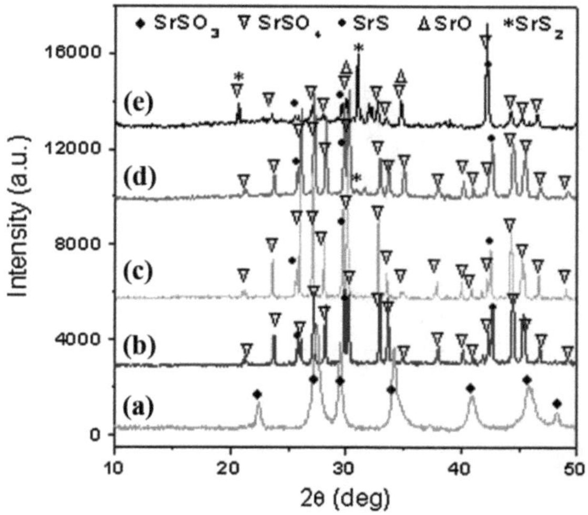

Fig. 4 XRD pattern of $SrSO_3$ (**a**) and reduction products of $SrSO_3$ at 700 °C (**b**), 900 °C (**c**), 1000 °C (**d**) and 1100 °C (**e**) for 7 min with N_2 protection

It demonstrates that the decomposition process of $SrSO_3$, which is wildly disputed, is the key step of reduction of $SrSO_4$. As a result, a newly prepared $SrSO_3$ was decomposed at different temperatures in the range of 973–1373 K with N_2 protection in experiments. The results, shown in Fig. 4, reveal that SrO cannot be formed until the temperature is above 1273 K.

A bulk of celestite ore was reduced at 1273 K, and the value of $SrSO_4$/SrS molar ratio in the reduction product on its surface was determined (in Fig. 5a1 and a2). The plots of $SrSO_4$/SrS molar ratio, shown in Fig. 5b, illustrate that $SrSO_4$ and SrS coexist in the reduction product, and the value of $SrSO_4$/SrS molar ratio is stable.

In summary, the reduction process of $SrSO_4$ is according to mechanism 3, namely reduction–decomposition mechanism, as follows: $SrSO_4$ converts to $SrSO_3$ at first. The newly generated $SrSO_3$ can be decomposed into SrS and $SrSO_4$ at lower temperature, or SrO and SO_2 at higher temperature. SO_2 can convert to $S_2(g)$ with reducing agent. And newly generated $S_2(g)$ will convert SrO to SrS finally.

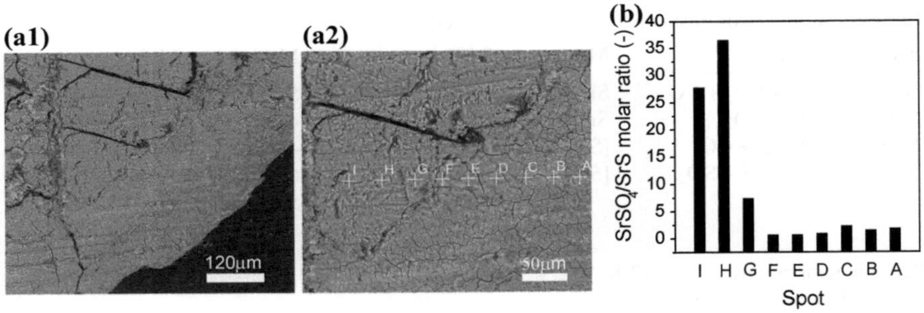

Fig. 5 SEM images of section of reduction product of celestite lump ore and graphite (**a2** is amplifying image of **a1**); the $SrSO_4$/SrS molar ratio of spot A–I in a2 (**b**)

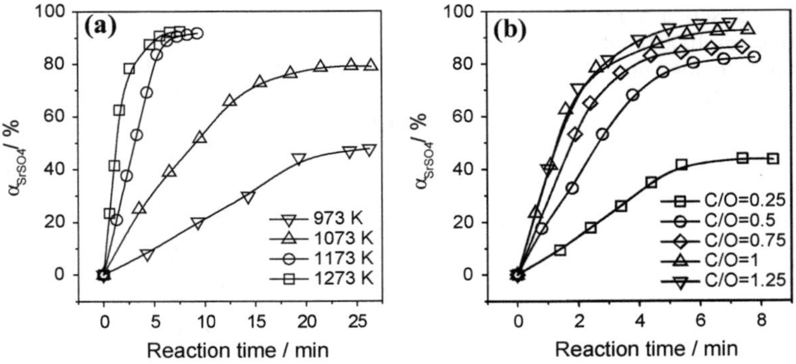

Fig. 6 Effect of temperature in the range of 973–1173 K (**a**) and C/O mole ratio in the range of 0.25–1.25 (**b**) on the reduction ratio of $SrSO_4$ in $SrSO_4$ pellet

The Effects of Reaction Temperature and C/O Molar Ratio on Reduction Ratio of $SrSO_4$

The effects of reduction temperature on α_{SrSO_4} are shown in Fig. 6a under the condition of C/O molar ratio set at 1.0. It can be seen that the value of α_{SrSO_4} is linearly increasing with increasing temperature in the early stage of the reaction. But the reaction rate of $SrSO_4$ decreases gradually with reaction time prolonging. It demonstrates that the reaction rate can be accelerated because of effective contact between $SrSO_4$ and graphite and lower diffusion resistance of reducing agents in the early stage of reaction. However, with the consumption of reducing agent, graphite is not in contact with $SrSO_4$, and the diffusion resistance raises. Besides, decreasing potential of reduction in reduction process of $SrSO_4$ causes the reaction rate to decline.

$SrSO_4$ carbon-containing pellet maintains high reaction rates in the range from 1073 to 1273 K, as shown in Fig. 6a. When the temperature is 1073 K, the reduction reaction of $SrSO_4$ can be completed within 12.5 min, and the final value of α_{SrSO_4}

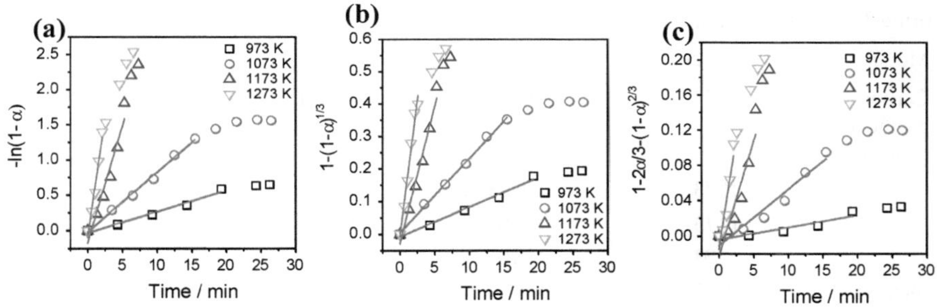

Fig. 7 Kinetic analysis of different limiting steps as carbon gasification (**a**), interface reaction (**b**) and diffusion (**c**)

reaches 65.6%. But as the reaction temperature increases to 1273 K, the reaction can be completed in 6.6 min, and the final value is 92.1%.

The effects of C/O molar ratio on α_{SrSO_4} are shown in Fig. 6b when the reaction temperature is set at 1273 K. It reveals that with increasing C/O molar ratio, the reduction rate of $SrSO_4$ increases generally. And the reduction rate of $SrSO_4$ stabilizes when the C/O molar ratio exceeds 1.0.

Kinetic Analysis of $SrSO_4$ in Carbon-Containing Pellet

The reduction process of $SrSO_4$ in carbon-containing pellet may be controlled by carbon gasification reaction, interfacial reaction or diffusion step, respectively. The classical theoretical calculation models for the limiting step of reaction are as follows [10]:

$$\text{Carbon gasification controlling:} \quad k_g t = -\ln(1-\alpha) \tag{19}$$

$$\text{Interfacial reaction controlling:} \quad k_i t = 1-(1-\alpha)^{1/3} \tag{20}$$

$$\text{Diffusion controlling:} \quad k_d t = 1-\frac{2}{3}\alpha-(1-\alpha)^{2/3} \tag{21}$$

Therefore, the data obtained from Fig. 6a are linearly fitted according to Eqs. (19)–(21), and the results are shown in Fig. 7 and Table 3. The value of k, slope of fitting line, represents the reaction rate constant of the reaction in the model, while the coefficient of determination R^2 represents the error between the fitted data and true values. From the analysis of standard deviation σ of R^2, the reduction process of $SrSO_4$ in carbon-containing pellet is limited by the interfacial reaction step. The apparent activation energy of the reaction is calculated by the method described in the literature [11], and the value is 105.4 kJ/mol.

Table 3 Simulation results of reduction process model for $SrSO_4$ pellet

Model	T (K)	k (min^{-1})	R^2	σ
Carbon gasification controlling	973	0.0297	0.9691	0.02225
	1073	0.0823	0.9891	
	1173	0.3303	0.9280	
	1273	0.6706	0.9693	
Interfacial reaction controlling	973	0.0090	0.9794	0.00929
	1073	0.0233	0.9956	
	1173	0.0842	0.9709	
	1273	0.1798	0.9883	
Diffusion controlling	973	0.0014	0.9267	0.03708
	1073	0.0056	0.8940	
	1173	0.0262	0.8253	
	1273	0.0503	0.8682	

Conclusions

(1) The result of thermodynamic analysis reveals that O_2 can inhibit $SrSO_4$ reduction. Reaction temperature and C/O molar ratio can affect the composition of $SrSO_4$ reduction product cooperatively. At higher temperature and relatively lower C/O molar ratio, SrO can be generated separately. The value of C/O molar ratio for the separate formation of SrO increases with increasing reaction temperature.
(2) The mechanism of reduction of $SrSO_4$ is to reduction–decomposition reaction.
(3) The reduction process of $SrSO_4$ in carbon-containing pellet is limited to the interfacial reaction step. The apparent activation energy of the reaction is 105.4 kJ/mol.

Acknowledgements The work was supported by Qinghai Province Science and Technology Support Program (2015-GX-108A). The authors also acknowledge Qinghai Zhongkeyuanhao strontium technology Co., Ltd. for its partial financial support of this investigation.

References

1. Dusza M, Stefanski M, Wozniak M, Hreniak D, Gerasymchuk Y, Marciniak L, Granek F, Strek W (2016) Luminescent Sr_2CeO_4 nanocrystals for applications in organic solar cells with conjugated polymers. J Lumin 169:857–861

2. Liu BB, Wang XH, Zhang RX, Li LT (2017) Energy storage properties of ultra fine-grained $Ba_{0.4}Sr_{0.6}TiO_3$-based ceramics sintered at low temperature. J Alloy Compd 691:619–623
3. Plaza M, Huang X, Ko JYP, Shen M, Simpson BH, Rodríguez-Lopez J, Ritzert NL, Letchworth-Weaver K, Guncele D, Schlom DG, Arias TA, Brock JD, Abrun HD (2016) Structure of the photo-catalytically active surface of $SrTiO_3$. J Am Chem Soc 138:7816–7819
4. Aydoğan S, Erdemoğlu M, Aras A, Uçar G, Özkan A (2006) Dissolution kinetics of celestite ($SrSO_4$) in HCl solution with $BaCl_2$. Hydrometallurgy 84:239–246
5. Duan DP, Han HL, Yuan P (2014) Experimental research on direct reduction of celestite by rotary hearth furnace process. J Salt Lake Res 22(1):27–31
6. Liu CL, Zhang L, Xu LJ, Su ZM, Xie TP, Wang Y (2014) Sr speciation in producing SrCO3 with celestite. Chin J Geochem 33:244–247
7. Chen W, Zhu Y, Yang HZ (1996) Study on producing strontium carbonate from celestite with reduction-dissociation method. J Kunming Univ Sci Technol 21(6):86–91
8. Zhang XM, Song XF, Sun Z, Li P, Yu JG (2012) Density functional theory study on the mechanism of calcium sulfate reductive decomposition by carbon monoxide. Ind Eng Chem Res 51(18):6563–6570
9. Ghardashkhani S, Cooper DA (1990) A thermogravimetric study of the reaction between sulfur dioxide and calcium oxide. Thermochim Acta 161(2):327–337
10. Ning GS, Li S, Zhang B, Liu CJ, Jiang MF (2017) Reduction kinetics of carbon-bearing pellets of red mud. Nonferrous Met (Extr Metall) 1:24–27
11. Guo XY, Tian QH, Yi Y (2016) Fundamental and technological study of high-arsenic dust by hydrometallurgical processes. Metallurgical Industry Press

2. Liu RP, Wang XH, Zhang [illegible] (201[illegible]) [illegible] properties of [illegible] supported $Ba_{0.5}Sr_{0.5}TiO_3$-based [illegible] at low temperature. [illegible]
3. Plaza M, Huang X, Ko JYP [illegible], Rodriguez-Lopez J, Batzel [illegible], Weaver K, Gumerlo [illegible] TA, Brock JD, Abruña HD (2016) [illegible] of the photoelectrochemically active surface [illegible] J Am Chem Soc 138(25):[illegible]
4. Atalogan S, Erdemoglu M, [illegible], Gökçe A (2005) [illegible] of a[illegible] $SrSO_4$ in HCl solution with BaCl[illegible] Hydrometallurgy 84:[illegible]40–[illegible]
5. Duan DP, Han HL, Yuan R (20[illegible]) [illegible] research on the direct [illegible] of [illegible] by rotary hearth furnace process. [illegible] Res 20[illegible]
6. Liu [illegible], Zhang L, Xu [illegible], Su Z[illegible], [illegible] Wang Y (2014) [illegible] speciation in [illegible] with [illegible]. Chin J Geochem [illegible]:[illegible]–247
7. Chen [illegible], Zhu [illegible], Yue [illegible] (1990) [illegible] determination of [illegible] with [illegible] reduction-[illegible] method. [illegible] Sci Technol 2(1):[illegible]–91
8. Zhang [illegible], Song [illegible], Sun Z, Li [illegible] (201[illegible]) [illegible] dynamic theory study on the mechanism of calcium sulfate [illegible] by carbon monoxide. Ind Eng Chem Res 51(18):[illegible]–65[illegible]
9. Chandrashekhar S, Cooper DA (19[illegible]) [illegible] study of the [illegible] of sulfur dioxide to calcium sulfide. Thermochim Acta 1[illegible]
10. Ning GS, Li S, Zhang B, Liu CL, Jiang [illegible] (2017) Reduction kinetics of carbon-bearing pellets of red mud. Nonferrous Met [illegible]:[illegible]24–2[illegible]
11. Guo XY, Tian QH, [illegible] Y (20[illegible]) [illegible] and technological study on [illegible] by [illegible] non-metallurgical processes. Metallurgical Industry Press

Author Index

G. Azimi et al. (eds.), *Rare Metal Technology 2019*, The Minerals, Metals & Materials Series, https://doi.org/10.1007/978-3-030-05740-4

Subject Index

G. Azimi et al. (eds.), *Rare Metal Technology 2019*, The Minerals, Metals & Materials Series, https://doi.org/10.1007/978-3-030-05740-4

MIX
Papier aus verantwortungsvollen Quellen
Paper from responsible sources
FSC® C105338

If you have any concerns about our products,
you can contact us on
ProductSafety@springernature.com

In case Publisher is established outside the EU,
the EU authorized representative is:
Springer Nature Customer Service Center GmbH
Europaplatz 3, 69115 Heidelberg, Germany

Printed by Libri Plureos GmbH
in Hamburg, Germany